AF348470

Fish Diseases and Management

Dr A Gopalakannan is presently working as Assistant Professor in Department of Aquatic Animal Health, Management, Tamil Nadu Dr. M.G.R. Fisheries College and Research Institute, Tamil Nadu Dr. J. Jayalalithaa Fisheries University, Ponneri. He has completed his under and post graduate (Fisheries Biotechnology) in Fisheries Science. He has completed his Ph.D degree in Department of Biotechnology, Pondicherry University and has been worked as Intern Scientist in West Sea Mariculture Research Center, National Fisheries Research and Development Institute, South Korea for the period of two years. Later he was working as Assistant Professor in Department of Genetic Engineering, School of Bioengineering, SRM University. The area of his research work was quantification of shrimp virus (WSSV and HPV) by real time PCR and Microbial community analysis (16S rRNA sequencing) from biolfoc based shrimp culture, control of *Aeromonas hydrophila* in carp by using immunostimulants (chitin, chitosam and glucan from spent yeast) and probiotics. He has published 15 research papers in national and international journal and also presented more than 20 papers in the national and international conference. He has deposited 3 efficient probiotic bacterial sequences (16S rRNA sequence) and 57 heterotrophic bacterial sequences isolated from intensive shrimp culture system (Biofloc) to National Center for Biotechnological Information, USA. He is a member for many scientific societies. He has operated more than 10 state and central government and industrial research project as Co-principal investigator and principal investigator. He is handling courses related to Fisheries Biotechnology and Aquatic Animal Health Management to undergraduate and postgraduate students.

Dr A Uma is working as Professor and Head, Department of Aquatic Animal Health, Tamil Nadu Dr. M.G.R. Fisheries College and Research Institute, Tamil Nadu Dr. J. Jayalalithaa Fisheries University, Ponneri. She has 19 years of experience in teaching, research and extension. She has developed a E- module for a courses, extended technical guidance to about 52 students and has served as an expert in the doctoral committees of research students. Her research expertise include molecular and immuno diagnosis of diseases in aquaculture, development of molecular and rapid diagnostics, disease resistance and innate immunity in aquatic organisms. Developed six technologies which include PCR and immuno based diagnostic kits for whitespot syndrome virus(WSSV), Monodon baculovirus(MBV) and Hepatopancreatic parvo virus (HPV) infecting shrimps and transferred two technologies to the field for the benefit of aquafarmers. Patent has been granted for her invention related to PCR diagnosis of WSSV in shrimps. Published about 49 research papers in national and international journals; full length papers in conferences; presented many research papers in national and international conferences and co-authored a book. She has trained shrimp hatchery and farm technicians on disease diagnosis and management techniques. She has undergone overseas training and is the recipient of Tamilnadu young woman scientist award of Tamilnadu State Council for Science and Technology, Young scientist fellowship award and K.Chidambaram Memorial annual award.

Dr S Felix, Ph.D., Vice Chancellor, Tamil Nadu Dr. J. Jayalalithaa Fisheries University, Nagapattinam is having experience of more than 35 years in the area of fisheries and aquaculture in teaching, research, extension and administration. He has organized more than 20 Nos. of seminar/conference/symposium at National and International levels. Currently he is also the President (2018-19) of World Aquaculture Society, Asian Pacific Chapter and Chairman, ICAR's BSMA Committee (2018-19). He has more than 60 research papers published in National and International journals. He has authored more than 10 books and 25 manuals. He is specialized in the area of advanced systems in aquaculture and aquariculture such as raceway, biofloc technology, RAS, etc,. He has operated around 20 externally funded research projects.

Fish Diseases and Management

by

A Gopalakannan

A Uma

S Felix

Tamil Nadu Dr. M.G.R. Fisheries College & Research Institute
Tamil Nadu Dr. J. Jayalalithaa Fisheries University
Ponneri, Tiruvallur District, Tamil Nadu

DAYA PUBLISHING HOUSE®
A Division of
ASTRAL INTERNATIONAL PVT. LTD.
New Delhi – 110 002

© 2018 UNIVERSITY
ISBN: 9789388173216 (Int. Edition)

Publisher's Note:

Every possible effort has been made to ensure that the information contained in this book is accurate at the time of going to press, and the publisher and author cannot accept responsibility for any errors or omissions, however caused. No responsibility for loss or damage occasioned to any person acting, or refraining from action, as a result of the material in this publication can be accepted by the editor, the publisher or the author. The Publisher is not associated with any product or vendor mentioned in the book. The contents of this work are intended to further general scientific research, understanding and discussion only. Readers should consult with a specialist where appropriate.

Every effort has been made to trace the owners of copyright material used in this book, if any. The author and the publisher will be grateful for any omission brought to their notice for acknowledgement in the future editions of the book.

All Rights reserved under International Copyright Conventions. No part of this publication may be reproduced, stored in a retrieval system, or transmitted in any form or by any means, electronic, mechanical, photocopying, recording or otherwise without the prior written consent of the publisher and the copyright owner.

Published by : **Daya Publishing House®**
A Division of
Astral International Pvt. Ltd.
– ISO 9001:2015 Certified Company –
4736/23, Ansari Road, Darya Ganj
New Delhi-110 002
Ph. 011-43549197, 23278134
E-mail: info@astralint.com
Website: www.astralint.com

Laser Typesetting : **Classic Computer Services,** Delhi - 110 035

Printed at : **Neelam Graphics, Delhi - 110007**

डॉ. जे. के. जेना
उप महानिदेशक (मत्स्य विज्ञान)

Dr. J. K. Jena
Deputy Director General (Fisheries Science)

भारतीय कृषि अनुसंधान परिषद
कृषि अनुसंधान भवन-II, पूसा, नई दिल्ली 110 012
INDIAN COUNCIL OF AGRICULTURAL RESEARCH
KRISHI ANUSANDHAN BHAVAN-II, PUSA, NEW DELHI - 110 012

Ph. : 91-11-25846738 (O), Fax : 91-11-25841955
E-mail: ddgfs.icar@gov.in

Foreword

Aquaculture in recent years has witnessed significant growth in terms area coverage, intensification and also species diversification. Farming at higher stocking densities together with higher inputs provision is the order of the day, which often result in poor water quality, increased stress to the animals and consequent incidence of diseases. Disease has become the major threat for aquaculture development in the country today. Addressing health questions, therefore, has become a primary requirement for sustaining the growth of aquatic animal food production. Fish disease management focuses on the important diseases of commercially important cultivable finfishes and shellfishes, their diagnosis and control measures.

The present book entitled *'Fish Diseases and Management'* authored by Drs A. Gopalakannan, A. Uma and S. Felix deals with the general information and significance of diseases in relation to aquaculture; host pathogen interaction; diseases development in fish and shellfish; disease occurrence associated with the water quality parameters; aspects of bacterial, viral, fungal and parasitic pathogens and their relative significance to aquaculture; strategies for health management dealing with immune system of fish and shrimp; different types of vaccines and routes of application; and also eco-friendly control measures.

I am sure that the book will be a valuable resource for students, teachers, researchers, fish pathologists and aquaculture technicians concerned with the management of diseases of fish and shellfish.

I congratulate authors for their sincere efforts in bringing out this publication.

(J.K. Jena)

Message

Aquaculture is an important food producing sector. Production through aquaculture has reached a total of 106 million tonnes, 76.6 million tonnes of aquatic animals and 29.4 million tonnes of aquatic plants with an annual average growth rate of 6.6 percent since 1995. Disease is the major impeding factor that affects the development and expansion of aquaculture. Technological advancement has helped in the intensification of this sector with the introduction of many new species and application of modern technologies. This has not only increased the production but has also resulted in the disease outbreak both existing and emerging. Identification of disease is the first step towards its management. Hence, knowledge on the various diseases affecting the fishes cultured in India, their identification, methods of prevention and control will be helpful in the management of the diseases and to improve the production in aquaculture. As there are not many publications on fish diseases and their management particularly for the Indian Aquaculture this book on Fish diseases and management will be very useful for the farmers, researchers and students.

S. Felix
Vice Chancellor
Tamil Nadu Dr. J. Jayalalithaa Fisheries University

Preface

Aquaculture industry is rapidly expanding due to technological advancement in production techniques. Intensification tends to adversely affect the health of cultured fish resulting in various diseases caused by virus , bacteria, fungi and parasites and also the emergence of new pathogens. This book contains 6 major chapters which deals with the commercially important diseases in fish and shell fish of bacterial, viral, fungal and parasitic origin, their symptoms, diagnosis and control measures and the eco-friendly management measures to be adopted in aquaculture.

This book will be a valuable resource for researchers, students, diagnosticians, fish, shrimp hatchery and farm technicians, consultants, fish pathologists and microbiologists concerned with the management of diseases of fish and shellfish.

We are greatly indebted to Dr. S. Felix, Vice-chancellor, Tamil Nadu Dr. J. Jayalalithaa Fisheries University, Nagapattinam who has accorded permission for publishing this book. We sincerely thank Dr. B. Ahilan, Dean, FC&RI, Ponneri for his encouragement. We profusely thank the Indian Council of Agricultural Research, New Delhi for the financial assistance.

We would welcome and very much appreciate constructive criticism and comments for the improvement of this manual in the future edition.

Authors

Contents

Significance of Fish Diseases in Relation to Aquaculture; Host, Pathogen and Environment Interaction in the Disease Development in Fish and Shellfish and Role of Stress in Disease Development; Role of Water Quality in Fish Diseases.

INFECTIOUS DISEASES OF FISH AND SHELLFISH

A. Bacterial Diseases of Finfish; 1. Disease Caused by Gram Negative Bacterial Pathogens; a.1. Infection Caused Due to Bacteria of Aeromonadaceae; b. Diseases Caused by Enterobacteriaceae; c. Disease Caused by Bacterial Pathogens of Cytophagaceae; 2. Disease Caused by Aerobic Gram-positive Rods and Cocci; a.1 Bacterial Kidney Diseases (BKD); a.2 Streptococcosis; a.3 Mycobacteriosis; a.4 Nocardiosis; 3. Disease Caused by Anaerobic Bacteria; B. Bacterial Diseases of Shellfish; a.1 Vibriosis; Vibriosis in Hatcheries; a.2 Bacterial Shell Disease; Brown or Black Spot Disease; a.3 Filamentous Bacterial Infection; a.4 Mycobacterial Infections; a.5 Rickettsial and Chlamydial Diseases.

Aquaculture; c. Immunostimulants; Concept of Immunostimulants; Application of Immunostimulants Used in Fish and Shrimp Synthetic Chemicals; Biological Substances; Polysaccharides; Animal and Plant Extracts; Nutritional Factors; Hormones; Algal Derivatives; Immunological System in Fish; Lymphocytes; Granulocytes; Macrophages; Immunostimulation of Non-specific Defense Mechanisms; Method of Administration; Timing of Administration; Mode of Action; Detection of Immunostimulation; Attributes of Immunostimulants; Efficacy and Limitation of Immunostimulants; Immunostimulants in Aquaculture Health Management; D. Probiotics; Selection of Probiotics; Selection of Candidate Species; Evaluation of Pathogenicity of Selected Strains; In vivo Evaluation of Potential Probiotic Effects on the Host; Experimental Infections–In vivo Antagonism Tests; Mode of Action; a. Competitive Exclusion; ii. Competition for Adhesion Site; iii. Antibiotics Production; iv. Siderophores Production; Characteristics of Good Probiotics; Constraints to Probiotics in Aquaculture; Probiotics Significance in Aquaculture; Improvement in Water Qualities; As Growth Promoters; For Disease Prevention; Source of Nutrients and Enzymatic Contribution to Digestion; Stress Tolerance; Enhancement of the Immune Response; Antiviral Effect; Effect on Reproduction of Aquatic Species; Probiotics in Aquaculture Management; Application in Feed; Direct Application to Pond Water; Application through Injection; Bioremediators and Other Prophylactic Measures Bioremediators; Other Prophylactic Measures.

Chapter 1

Introduction

Significance of Fish Diseases in Relation to Aquaculture

Aquaculture has developed into a science and an industry following the declining trend of capture fisheries sector in the last two decades. In order to increase profitability, the culture practices has turned from semi-intensive to intensive systems, leading to indiscriminate increasing of stocking density resulting in increased stress to the animals and consequent incidence of diseases. Disease has now become a primary constraint to aquaculture growth and addressing health questions has therefore become an urgent requirement for sustaining the growth of aquatic animal food production.

Host, Pathogen and Environment Interaction in the Disease Development in Fish and Shellfish and Role of Stress in Disease Development

As could be seen from the classical description of Snieszko in 1970, the incidence of disease is due to the breakdown of the delicate balance between host, pathogen and environment. The host in many cases could resist invasion of most of the pathogens if it is healthy and free from stress. In many of the aquaculture systems however, such a situation rarely exist, as there are a multitude of variables that need to be kept under check for a balanced and optimum requirement of the cultured species. The aquatic environment by itself is highly dynamic and the body functions of fishes which are poikilothermic are controlled by temperature and other water quality parameters. Fish has to continuously adjust to the changing environmental parameters, else it leads to low productivity, reduced weight gain, reduced feed conversion, decreased immunity, reduced natural disease resistance, increase in infectious diseases, death and reduced profits for a commercial fish farmer.

Irresponsible use of chemical disinfectants and antibiotics is also a potential environmental hazard. In fish, stress is usually related to stocking density, environmental quality, handling or transport. Additionally, intensive culture practices with poorly controlled feed use and waste production adversely affect the environment. Variations in the environmental and biological parameters from the optimum range introduce stress to the animals under culture, which has a direct bearing on their immune system. In fish, stress response is initiated by adaptive changes due to the stimulation of hypothalamus-pituitary-inter renal (HPI) axis and consequent production of corticosteroids. Although these stress hormones help the animal in its effort to regain internal homeostasis, they significantly lower the defence mechanisms due to their immuno suppressive nature.

Role of Water Quality in Fish Diseases

Maintenance of good water quality by making inlet water free from toxic substances and pathogens is one of the integral parts of the health management in aquaculture. It has been advised that the shrimp farmers maintain a reservoir pond for pre-treating the inlet water. Water quality deterioration if noticed in a farm has to be attended immediately. Increasing aeration, reduction of feeding rates, control of phytoplankton bloom and water flow management are the important steps that are to be undertaken without any delay for abating low dissolved oxygen content. Water exchange during periods of poor water quality also helps to regain the health status of the cultured fish.

Uneaten food, faecal waste, dead plant matter will increase the organic load. Addition of calcium hydroxide or calcium carbonate enhances respiration of acidic soils. For optimum oxidation and respiration of organic matter, soil moisture has to be between 12 per cent to 20 per cent and at pH 7.5 to 8.0. Low or fluctuating pH, low alkaline or low hardness condition can be corrected by adding agricultural lime in recirculation or flow through systems by dipping lime into the water or passing water through a bed of oyster shells. Introduction of biofilter has been found to be very much essential in closed circulatory system to keep up the water quality. It is also advisable to sterilize the incoming water to hatcheries for fear of entry of potent viral pathogens, which could often be accomplished by the ozone treatment.

Water quality criteria for optimum fish health management of warm water and cold water species of fish (mg/l, except pH)

Characteristic	Cold Water	Warm Water
O_2	5– Saturation	5– Saturation
pH	6.5–8	6.5–9
Ammonia (un-ionised)	0–0.125	0–0.02
Calcium	4–160	10–160

Characteristic	Cold Water	Warm Water
CO_2	0–10	0–15
H_2S	0–0.002	0–0.002
Iron (total)	0–0.15	0–0.5
Manganese	0–0.01	0–0.01
Nitrate	0–3.0	0–3.0
Phosphorus	0.01–3.0	0.01–3.0
Zinc	0–0.05	0–0.05
Total Hardness ($CaCO_3$)	10 – 400	10–200
Total alkalinity ($CaCO_3$)	10 – 400	10–400
Nitrogen (gas saturation)	<100 per cent	<100 per cent
Total solids	0–80	1.–500

A disease is the sum of the abnormal phenomena displayed by a group of living organisms in association with a specified common characteristic or set of characteristics by which they differ from the norm of their species in such a way as to place them at a biological disadvantage. The reason for the occurrence of disease out breaks is varied, representing complex interactions between the host and the disease-causing situation. Some bacteria are primary pathogens while others are opportunistic, (saprophytic) water borne bacteria that would colonize the exposed tissue. The opportunistic pathogen can cause a disease when the resistance of the fish reduces (in an adverse physiological state) as in an environmental stress and an increase in number and/or virulence of the pathogens. A disease may even result from the synergistic interaction of two or more organisms. Within fish farms, outbreaks of disease may begin suddenly, progress rapidly usually with high mortalities and disappear with equal speed (acute diseases) or develop quite slowly with less severity but persist for greater periods (chronic out breaks).

According to Kinne (1980) four major groups of diseases may be identified in terms of epizootiology. These include 'sporadic disease' which occur sporadically in comparatively few members of a population; 'epizootics' which are large scale out breaks of communicable animal disease occurring temporarily within limited geographical areas; panzootics, which occur in large areas; and enzootics which persist or re-occur as low level out breaks in certain areas.

The importance of diseases in aquaculture is dependent on the type of culture system employed. There are three basic types of culture systems in use in modern day fish or shrimp farming:

1. Intensive culture system, in which animals are raised in high density, intensively managed tanks and ponds.

2. Semi-intensive systems in which fish or shrimps are raised in moderate densities in ponds, cages or tanks with some management of the system practised.

3. Extensive systems where they are grown in low intensity ponds or pens in natural bodies of water and where little or no management is practised.

Hatcheries are a part of the intensive or semi intensive culture systems. As the stocking density in a culture system increases, the possibility of disease occurs. Many bacteria causing diseases in aquaculture are a part of their normal microflora. They establish lethal influence as a result of other primary conditions such as nutritional disorders, extreme environmental stress and wounds. In the ponds when the waste materials like uneaten feed, and metabolites of animal increases, bacteria multiply in large numbers. When the load of bacteria in the medium is high and the animals are stressed due to any of the conditions, opportunistic bacteria can cause diseases.

INFECTIOUS DISEASES OF FISH AND SHELLFISH

Chapter 2

Bacterial Diseases

A. Bacterial Diseases of Finfish

1. Disease Caused by Gram Negative Bacterial Pathogens

a.1. Infection Caused Due to Bacteria of Aeromonadaceae

There are different species of bacteria belonging to the genus *Aeromonas*, capable of causing disease in fish. The important species are *Aeromonas salmonicida* - aetiological agent of furunculosis in salmonids and *A. hydrophila* which mainly affect tropical fishes. They are non-sporing, facultatively anaerobic rods with rounded ends and resistant to the vibriostat compound (2,4 diamino 6, 7 diisopropyl pteridine). Optimum growth temperature is 22-28°C and some do not grow at 35°C. They occur widely in freshwater and sewage. Some are pathogenic to fish but rarely to man.

a.2. *Aeromonas salmonicida*

Aeromonas salmonicida is the causative agent of furunculosis in salmonids and related diseases in other species. On commonly used culture media, it often produces a diffusible brown pigment which helps in its identification. *A. salmonicida* is non-motile and an obligate pathogen of fish with limited survival outside the host. The name furunculosis is given because of its sub acute or chronic form, recognized by the presence of lesions resembling boils *i.e.*, furnuncles in the musculature. The sub acute or chronic form of furunculosis, in older fish, is characterized by lethargy, slight exophthalmia, blood shot fins, bloody discharge from the *nares* and vents, and multiple hemorrhages in the muscle and other tissues. Internally, hemorrhaging in the liver, swelling of the spleen, and kidney necrosis may occur.

This bacteria can cause

(a) Furunculosis of salmonids

(b) Goldfish ulcer disease

(c) Carp erythrodermatitis

(d) Trout ulcer disease.

Additionally, several other species of Aeromonas, including: *A. hydrophila, A. formicans, A. liquefaciens,* and *A. hydrophila* complex are capable of causing a disease known as "Motile aeromonas septicemia" or "Bacterial hemorrhagic septicemia".

a.3. Furunculosis of Salmonids

Typical furunculosis in salmonids may occur in one of several forms:

Peracute Form in Fingerlings

These fish usually have a dark discoloration and die rapidly without any other premonitory signs. The gross lesions may resemble those observed in the acute form of the disease.

Acute Form

Premonitory signs of anorexia occur 2-3 days prior to death. Gross lesions include hemorrhage of the liver and splenomegaly.

Sub-acute

This form has a much slower onset of clinical signs with petechial hemorrhages being observed in the skin and around the fins. Fish exhibit focal discolorations and anorexia and die approximately 4-6 days after the onset of clinical signs. Gross lesions include typical "furuncles" as well as internal lesions listed above for the acute form.

Chronic

This form is observed in fish which survive the subacute form and is manifested by healing of the furuncles and scarring. Affected and recovered fish may exhibit poor weight gain and be focally discolored.

Histologic lesions observed in the peracute and acute form will demonstrate necrosis, hemorrhage and bacterial microcolonies within affected organs. Furuncles classically are described a dark, raised tumefaction involving the skin, subcutis and underlying skeletal musculature. These lesions will ulcerate and drain a serosanguinous fluid. These lesions develop from localization of hematogenous bacteria in the muscle or skin, not from an external skin leison. The lesion histologically is characterized by marked necrosis of the skin, subcutis and skeletal muscle with mild to minimal acute inflammatory infiltrates in the acute stage. In the chronic form, scarring will be characterized by the replacement of muscle tissue with fibrous connective tissue. Culture of acute lesions will often recover the organism in pure culture.

a.4. Cutaneous Ulcerative Disease of Goldfish

This disease, although it is most common in goldfish, also affects other non-salmonid fishes. It is caused by *Aeromonas salmonicida,* and the disease is also referred to as "furunculosis". The skin lesions range from whitish discolorations to shallow hemorrhagic ulcers to deep lesions which may expose underlying muscle or bone. These lesions can become secondarily infected with fungi, protozoa, or other bacterial agents. Fish may exhibit hemorrhage on the body as well as the base of the fins.

Diagnosis of this disease is dependent upon culture of the etiologic agent from the lesions. Histologically, the lesions have a mild to moderate primarily mononuclear inflammatory infiltrate. Large numbers of bacterial microcolonies are observed in many of the lesions.

a.5. Motile Aeromonas Septicemia (MAS)

This is the third manifestation of disease caused by *Aeromonas* sp. This is probably the most common bacterial disease of freshwater fish. This disease has been associated with several members of the genus Aeromonas, including *A. hydrophila, A. sobria, A. caviae, A. schuberti,* and *A.veronii.*

Clinical signs of motile aeromonas septicemia range from sudden death with high morbidity in peracute cases to superficial to deep skin lesions. Skin lesions include variously sized areas of hemorrhage and necrosis and the base of the fins. These lesions may progress to reddish to gray ulcerations with necrosis of the underlying musculature. Ulcers may be observed in conjunction with a hemorrhagic septicemia which can produce non-specific lesions and clinical signs of exophthalmos, ascites, visceral petechiation and a hemorrhagic and swollen lower intestine and vent. Anorexia and cutaneous discoloration are also observed with the septicemia.

Histopathological lesions include acute-to-chronic dermatitis and myositis. With the septicemia, there may be depletion and necrosis of the renal and splenic hematopoietic tissue, as well as necrosis in the intestinal mucosa, heart, liver, pancreas and gonad.

Definitive diagnosis of MAS requires biochemical identification of the suspected bacterial agent within the target tissues, with attendent clinical signs and lesions. The kidney is probably the best target tissue for culture, however, clinical lesions should also be cultured.

Control measures include both prophylaxis and therapeutics. The several methods of disease control advocated include adequate husbandry practices, such as maintenance of good water quality, disinfection of fish farm where outbreaks occur and routine disinfection policies for eggs upon arrival at receiving sites. Stocking of fish farms with disease-resistant strains of fish and the development of effective vaccines also improve the health of fishes. In case of therapy, control measures have centered around use of antimicrobial compounds.

Chemotherapy used for *A. salmonicida* include sulphonamides (mainly sulphamerazine), chloramphenicol, oxytetracycline and furazolidone. Sulphamerazine was given orally with feed at a dose of 22 g of drug/100 kg of fish/day. Chloramphenicol and oxytetracycline were dosed 5-7 g/100 kg of fish/day. The effectiveness of chemotherapy in the treatment of fish diseases largely reflects the method of administration. The various methods advocated are the oral method by means of medicated food, bath and dip procedures and injection.

Vaccines have been used to prevent the furunculosis and are successful. Commercial vaccines against this pathogen are available. Major administration route is through intraperitoneal injection.

a.6. *Aeromonas hydrophila*

Aeromonas hydrophila is a motile aeromonad recognized as a causative agent of haemorrhagic septicemia. *A. hydrophila* has been recovered as a pathogen from a wide variety of freshwater fish species and occasionally from marine fish. It causes distinct pathological conditions, including tail/fin rot and haemorrhagic septicemias. This organism is also found associated with other pathogens. Haemorrhagic septicemia (motile aeromonas septicemia) is characterized by small surface lesions, local haemorrhages particularly in the gills and vent, ulcers, abscesses, exophthalmia and abdominal distension. Internally, there may be accumulation of ascitic fluid, anemia and damage to the organs, notably kidney and liver.

Redsore disease in seabass has also been attributed to *A. hydrophila.* This disease can reach epizootic proportions and is characterized by erosion of the scales and pin-prick haemorrhages which may cover up to 75 per cent of the body surface with a high mortality rate.

This organism is present in freshwater and can occur in the natural ecosystems of all the fish farms such as cultured cyprinids or catfishes. In most cases, a stress mediated disease condition has been observed. Seasonal outbreaks of diseases due to this are usually associated with primary viral or parasitic infection. Such secondary invasions with motile aeromonads also characterize a wide range of other diseases such as epizootic ulcerative syndrome (EUS), furunculosis of salmonids, red-sore disease of large mouth bass and many parasite condition of tropical farmed fishes. *A. hydrophila, A. caviae* and *A. sobria* have all been isolated from fish with clinical signs and bacterial septicaemia, although *A. hydrophila* is by far the most frequently associated.

Control of *A. hydrophila* infection is ideally linked to the control of underlying factors which have facilitated its invasion of the host. For chemotherapy, chloramphenicol and nifurpirinol, oxytetracycline and sulpham erazine have been used. A major problem with antibiotic treatment of *A. hydrophila* as with most bacterial infections is the development of plasmid related antibiotic resistance.

Vaccination is an alternative method of prophylaxis, as in diseases such as vibriosis and enteric red mouth. Since the motile aeromonads show great antigenic diversity, a polyvalent vaccine would be necessary for combating this disease. Since they are invariably associated with the invasion of a pathogenic strain into a stressed population, control at least in farmed populations has to be aimed primarily at reducing the level of possible stresses.

a.7. Pasteurellosis

It is the bacterial disease caused by *Pasteurella piscicida* first reported from white perch in U.S.A. during the year of 1963. The affected fish shows prominent white granules in the kidneys and spleen. Pasteurellosis is one of the most important bacterial disease among young cultured marine fish in Japan.

Clinical signs and pathology varies with the form of disease. In acute cases, the body will be darkened, and internally granulomatous – like deposits may develop in the kidney and spleen. Hence, the name pesudotuberculosis is given. In histopathological studies, the tubercle like lesions composed of masses of bacteria, epithelial cells and fibroblasts can be found.

Pasteurella piscicida is a Gram negative short rod (0.6 – 1.2 x 0.8 – 2.6 mm) which exhibits characteristic bipolar staining. It is non-motile, non-capsulated and non-spore forming. Serological tests such as slide agglutination using specific antigen or direct and indirect fluorescent antibody techniques can be used for definitive diagnosis.

When the water temperature increases to 25°C or more or when the salinity is reduced by long term rain, pseudotuberculosis caused by *P. piscicida* appears in fish farms in Japan. *P. piscicida* does not survive for more than 4-5 days in marine waters.

Avoidance of over crowding and feed management may prevent out-breaks of this disease. Ampicillin, amoxicillin, novobiocin, chloramphenicol, florfenicol, oxolionic acid, flumequin and sodium nifurstyrenate are used to control bacterial pseudotuberculosis.

a.8. *Pseudomonas*

Pseudomonas spp. are frequently associated with fish and are found on eggs, the skin, gills and intestine. The general bacterial flora of fish including *Pseudomonas, Cytophaga, Flavobacterium, Micrococcus* and *Acinetobacter* reflect the microbial population of the aquatic habitat and is influenced by factors such as bacterial load in the water and salinity. Among the pseudomonads, *P.anguilliseptica* is described as the causative agent of haemorrhagic bacteraemia of cultured cells. The organism can be isolated from the liver, spleen, kidney, heart and colonies can be formed on nutrient agar after 72 h incubation at 25°C.

Pseudomonas fluorescens produces the pseudomonas septicemia, which is a haemorrhagic condition of fish usually associated with stress or improper management. Many cultivable and aquarium fishes are affected by the disease.

P. fluorescens can be isolated from the kidney lesions of the affected fish. Probably all species of fish are vulnerable to pseudomonas septicemia under adverse environmental condition or when compromised by other factors.

b. Diseases Caused by Enterobacteriaceae

b.1 Edwardsiellosis

This disease is also known as emphysematous putrefactive disease (EPD) and is caused by *Edwardsiella tarda*. It is a gram-negative motile rod. Normally, this disease occurs in the summer months with an optimum temperature of 30°C. The source of *E. tarda* is presumably intestinal contents of carrier animals.

Clinically, lesions are initially observed as 3 to 5 mm red cutaneous foci on the flanks and caudal peduncle. They are caused from fistulas originating deep in the skeletal muscle. There is petechiation and malodorous liquefactive necrosis of the viscera with fibrinous peritonitis. Catfish affected with this disease will continue to eat even if they are severely affected. There may be posterior paresis in the later stages of the disease.

Definitive diagnosis is based on identification of the bacterium within the lesions and the attendent clinical findings. A fluorescent antibody test is available for identification of the bacterial agent, using kidney as the target tissue.

Enteric septicemia of catfish is caused by *E. ictaluri*. It is a highly contagious systemic disease. This disease is also called as "the hole in the head" due to the presence of an open lesion appearing in the skull of some chronically ill fishes. *E. ictaluri* is a typical member of the Enterobacteriaceae; gram negative, cytochrome oxidase-negative, rod and weakly motile at 25°C–30°C. The diseased fish hangs listlessly at the surface with a head up tail down position, sometimes rapidly increased spinning, usually followed by death.

Externally petechial hemorrhage or inflammation occurs in the skin under the jaw on the opercula and belly. These lesions often become so numerous that the skin appears bright red. Inflammation and haemorrhage also occurs at the base of all fins. In chronically infected fish, an open lesion will develop between the frontal bones of the skull, posterior to or between the eyes. Affected fish have pale gills, exophthalmia and often an enlarged abdomen. Internally, the body cavity may be filled with a cloudy, bloody or rarely a clear yellow fluid. The kidney and spleen are enlarged and spleen is dark red. Inflammation occurs in adipose tissue, peritoneum and intestine and the liver is either pale or riddled with congestion.

The only control is through the use of antibiotics incorporated into the feed. Oxytetracycline and a potentiated sulfonamide are the only drugs currently in use. Since the appetite of the fish in lost after some time the medicated feed should be given early in the epizootic.

b.2 Yersiniosis or Enteric Redmouth (ERM)

Yersinia ruckerii is the causative agent of enteric redmouth disease. This is an important pathogen of salmonids. All ages of fish are susceptible but primarily affects market sized fish. Severity of disease depends on strain virulence and degree of environmental stress. Lower mortality is observed under low temperature.

Early stages of ERM resemble aeromonad and *Vibrio* infections. There is darkening of the dorsum, anorexia and lethargy. Internal lesions are typical septicemia including visceral petechiation, splenomegaly and necrosis of the intestinal mucosa with a catarrhal exudate. With chronic disease there is also abdominal distension, unilateral or bilateral exophthalmia and hyperemia (blood spot). Due to ophthalmic lesions, which causes blindness, leading to lack of melanin pigment control and darkening. Histologically there is bacterial colonization of well-vascularized tissues, causing hemorrhage of gills, kidney, liver, spleen and heart, as well as muscle. This leads to necrosis of haematopoietic tissue causing anaemia. There is also necrosis and sloughing of gastrointestinal tract. Definitive diagnosis of ERM requires identification of the bacterium in target tissue with clinical signs. Kidney is the best organ for isolation during epidemics.

For control of this disease, oxytetracycline and ormetoprim – sulfadimethoxine are used. Many isolates are found susceptible to oxolinc acid. Disease carriers are the most important source of infection, especially when stressed. Keeping the water supply free of carrier fish is the best method of control. Maintenance of good sanitation and minimum stress should be assured to reduce reoccurence of carriers. Keep fish-eating birds and mammals away from culture facilities since many can transport the bacterium in their intestines. Commercial bacterins offer good protection and are important in managing population at risk for ERM.

b.3 Enteric Septicemia of Catfish (ESC)

This is probably the most important bacterial disease of catfish, Ictalurids such as brown bullhead, blue and white catfish. This is a markedly seasonal disease, with outbreaks occurring when water temperatures are in the range of 24–28°C.

Acute Form

Bacteria are ingested into the gastrointestinal tract and enter the bloodstream through the intestine with subsequent colonization of internal organs with resultant necrosis of these organs. There is usually a very high mortality and in some cases, very few premonitory clinical signs are observed.

Clinical signs may include corkscrew spiral swimming, abdominal distention, exophthalmos, or pale gills. Petechial hemorrhages may be

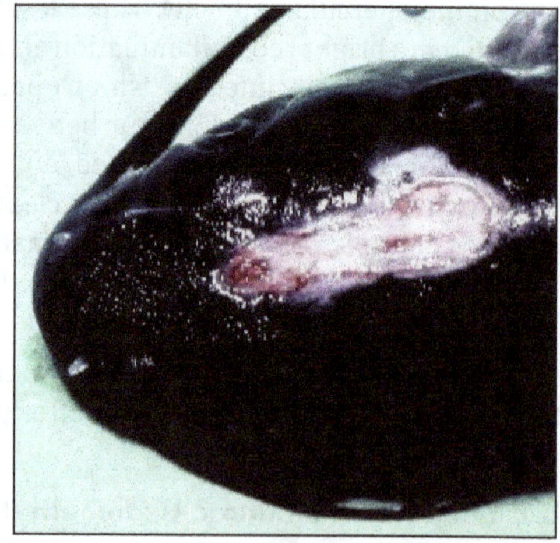

observed on dark areas of the skin as small (1 to 3 mm) depigmented foci (called "false spots"). Internally, the peritoneal cavity contains a bloody or clear fluid, hemorrhage and necrosis of the liver, as well as splenic and renal hypertrophy.

Chronic Form

Bacteria enter the catfish *via* the nervous system, by invading the olfactory organ *via* the nasal opening and migrate up the olfactory nerve to the brain, where the infection spreads from the meninges to the skull and finally to the skin, forming the classic "hole-in-the-head" lesion. This is a raised or open ulcer on the frontal bone of the skull.

Histologically, enteritis, hepatitis, myositis, and interstitial nephritis begin as acute lesions and develop into chronic foci. Fish with the nervous form develop neuritis of the olfactory nerve with meningoencephalitis developing in the late stages. Definitive diagnosis of ESC requires identification of the gram negative bacteria in target tissues, with attendant clinical signs.

b.4 Vibriosis

The genus vibrio contains gram negative, straight or curved rods which are non spore forming and motile by monotrichous or multitrichous sheathed polar flagella. They are common in aquatic habitats particularly in marine and estuarine habitats. Several species are pathogenic for man as well as marine animals. Metabolism is aerobic or facultative anaerobic and carbohydrates are fermented with the production of acid, but not gas.

V. alginolyticus, V. anguillarum, V. ordali V. salmonicida, V. vulnificus and V. damsela are the major fish pathogenic vibrios. In most of the cases anemia in a feature of vibriosis and was due to haemolysis. The exotoxins produced by these bacteria are proteins and are enzymic in nature. The pathogenic strains have well developed iron sequestering mechanism based on secretion of a siderophore, which induces separation of plasma, and tissue iron from its transferin or ferritin binding protein.

Whenever marine fishes are stressed or traumatised, in wild fish and farmed fish, vibrios are able to produce infection which begins as local hemorrhagic ulcers on the mouth or skin surface, or else local necrotic lesions in the muscle, the orbit or along the edge of the fins.

b.4.a. V. anguillarum

When the temperature is high with high salinities a condition called red pest out breaks occurs in eels similar to acute vibriosis. Vibriosis occurs in warm weather, when the stocking densities are high and salinities and organic loads are also high. In young fish, mortalities can be 50 per cent or higher. In older fish, infected fish do not feed and normal appearing fish will have necrotic lesions in muscle mass. Signs of the disease are anorexia, with darkening of whole fish or particular areas. In the acute stage, fish have dark skin swellings ulcerating to release a red effusion of necrotic tissue containing large number of infective bacteria. Chronically infected fish have large granulating lesions deep in the muscle. Gills are pale and fibrinous adhesions occur between visceral and parietal peritoneum. The eye may get infected and ulcerated.

Histologically, they show cardiac myopathy, renal and splenic necrosis of the erythroid haematopoietic tissue and periorbital oedema in per acute cases. In acute cases, skin lesions extend deep into the muscle. Severe necrosis of muscle with internal ulceration occurs. In the liver, there is focal necrosis and in the spleen and in kidney, massive depletion and necrosis of hematopoietic elements.

b.4.b. *Vibrio salmonicida*

V. salmonicida is specifically responsible for the condition of Atlantic salmon coldwater vibriosis or Hitra disease. Clinical signs of hitra disease include disorganized swimming. Fish may however die in pre acute phase without any obvious clinical signs. Common external signs are pallor of the gills, haemorrhage on the fin base, redness and swelling with occasional prolapse at the rectum and petechial hemorrhage on the ventral abdominal wall. There is hemorrhage on the swim bladder, within the abdominal fat and on the hepatic and other visceral peritoneal surfaces and a pale gray spleen.

It is best achieved by maintenance of water quality, good husbandry and low stocking densities. Killed vaccines against *V. anguillarum* have been available since 1980. All vaccines contain heat stable lipopolysaccharides derived from cell walls. The administration is by injection, oral and immersion method.

c. Disease Caused by Bacterial Pathogens of Cytophagaceae

c.1 Columnaris Disease

Columnaris, is a common bacterial disease that affects the skin or gills of freshwater fish and is caused most commonly by *Flexibacter columnaris*. This disease is also known by a wide variety of synonyms including the following: mxyobacterial disease, peduncle disease, saddleback, fin rot, cotton wool disease, and black patch necrosis. This bacteria is usually pathogenic at temperatures greater than 59°F. Both mortality and acuteness of the disease will increase at higher water temperatures. Virulence mechanisms associated with this disease are not well understood, however, the mineral content of the water is thought to be important, since this bacteria has been shown to be less pathogenic in soft water as compared to hard water. Other risk factors include physical injury, low dissolved oxygen, organic pollution and high nitrite levels.

This is primarily an epithelial disease, *i.e.*, it causes erosions and necrosis of the skin and gills which may become systemic. It often presents as whitish plaques that may have a red peripheral zone on the head or back (hence the name saddleback) and/or the fins (hence, fin rot) and especially the caudal fin (hence, peduncle disease). Fragments of the fin rays may remain after the epithelium has sloughed, leaving a ragged appearance. Lesions rapidly progress to ulcers, which may be yellow or orange due to masses of pigmented bacteria. Ulcerations spread by radial expansion and may penetrate into deeper tissues, producing a septicemia. Gill infections are less common but more serious. Columnaris begins at the tips of the lamellae and causes a progressive necrosis that may extend to the base of the gill arch. Definitive diagnosis is dependent upon the isolation of the bacterial agent in the presence of attending clinical lesions. It should be noted that a presumptive identification of *Flexibacter columnaris* can be made by examination of wet mounts and observation of long thin bacterial rods which a characteristic flexing or gliding motion. In addition to *Flexibacter columnaris*, other bacterial agents which have been implicated in this disease include: *Flexibacter psychrophila* as well as Cytophaga and *Flavobacterium branchiophila*.

Since *F. columnaris* primary affects to external surfaces of fish, chemicals can be added directly to the water as a dip, flush, bath or indefinitely prolonged treatment. Sulphamerazine and oxytetracycline are administrated theraputically in feed in a 2 stage regime, 220 mg/kg/day for 10 days followed by 50 to 75 mg/kg/day for 10 days.

c.2 Saltwater Columnaris Disease

This is caused by *Flexibacter maritimus* and was first reported from juvenile sea bream cultured in floating net cages in Japan. This organism has absolute requirement for seawater. Colonies can be isolated on cytophaga agar prepared with seawater. This is less infectious than *F. columnaris*. *F. maritimus* affects mainly the fishes less than 60 mm in body length and and affects mainly red sea bream. Treatment is to administer antibiotics such as oxytetracycline in the feed.

c.3 Bacterial Gill Disease (BGD)

Bacterial gill disease is a condition primarily associated with yellow pigmented filamentous, Gram negative bacteria of genera *Cytophaga*, *Flexibacter* and *Flavobacter.* Environmental conditions are thought to play some role in the aetiology of BGD. Affected fish become lethargic and anorexic, they tend to remain near the surface or inlet and may be observed flaring their opercula and coughing their respiratory rate is elevated and mucous secretion may increase. Gross pathological changes in mild or early stages include hyperemia, swelling of the primary lamellae and increased mucus secretion which traps debris. This may further invite secondary infection. BGD causing bacteria are called as myxobacterium in old literature. They are also called as yellow pigmented bacteria (YPB) and also referred as cytophaga-like bacteria (CLB).

For control, environmental quality should be maintained by avoiding over crowding, low dissolved oxygen, suspended solids and high ammonia levels. Bacteria affected are effectively removed by use of 1-5 per cent. NaCl for 1-2 minutes.

c.4 Finrot

It is applied to a wide range of conditions affecting many species. In addition to the presence of bacteria fin rot has been linked with traumatic damage, pollution and inappropriate nutrition. Various bacteria associated with finrot are CLB, aeromonads, pseudomonads, vibrios and unspecified genus.

2. Disease Caused by Aerobic Gram-positive Rods and Cocci

This is an important group of bacteria causing diseases to fishes. Important genera are *Coryneforms, Micrococcus, Mycobacterium, Nocardia, Planococcus, Rhodococcus, Renibacterium, Staphylococcus* and *Streptoviertieillium*.

a.1 Bacterial Kidney Diseases (BKD)

Renibacterium salmoninarum causes bacterial kidney diseases (BKD). This is also called as dee disease, corynebacterial kidney disease, salmonid kidney disease. The disease is found to occur to 13 species of salmonids in North America, South America and European countries, mainly occuring in the farmed fishes.

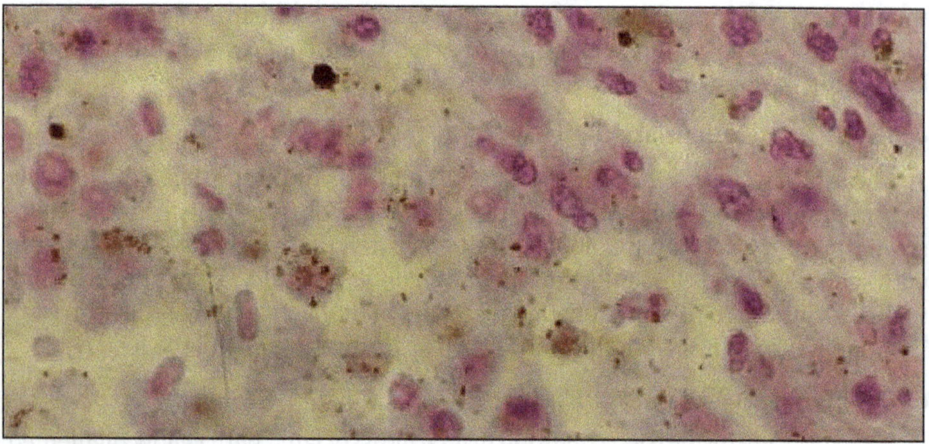

External signs of this disease include exophthalmia, swollen abdomen due to ascitic fluid, blood-filled blisters on the flank, and the presence of ulcers/abscesses. Initially lesions may develop in the kidney, liver, heart and spleen. The lesions contain a fluidy mass of leucocytes, bacteria and cellular debris.

a.2 Streptococcosis

Streptococcus is a large, complex genus of gram-positive spherical or oval bacteria less than 2mm in diameter. Outbreaks of streptococcal septicemia has been different in cultured freshwater fish in different parts of the world. Some host specificity exists and trouts suffer from heavy mortalities while carp and largemouth bass are generally not affected.

It varies with the species of affected fish. Erratic swimming, darkening of body color, unilateral or bilateral exophthalmia corneal opacity, hemorrhages on the opercular and the basis of the fins, and ulceration of body surface are the most common symptoms. The haemorrhagic lesions, gradually extend and ulcerate to release necrotic material. They are more superficial than the lesions of furunculosis or vibriosis. In many species, the eyes are affected with high frequency. There may be hyperemia of branchial vessels, infiltration with macrophages and necrosis leading to massive haemorrhage and death. The intestinal tract may be hyperemic with sloughing of mucosa.

The causative organism Streptococcus is present in seawater and mud around fish farms throughout the year, but in summer month their number increases. The presence of typical clinical signs and demonstration of Gram-positive cocci from internal organs constitutes a presumptive diagnosis. Confirmative diagnosis is by determining cultural characteristics and serological identification of the isolates. Fluorescent antibody technique is the most rapid and effective method for diagnosis of streptococcal infection. Mode of transmission of streptococcosis in fish is horizontal with infection occuring from direct contact with infected fish or contaminated fish food.

Pathogenicity is due to an exotoxin. Different strains of *Streptococcus* were lethal to different types of fishes like golden shiners, blue gill, green sunfish and american toads. "β–hemolytic" streptococcus spp. isolated from Ayu in Japan were very pathogenic for tilapia, yellow tail and sea bream.

Erythromycin is effective against streptococcal infections in cultured yellow tail and rainbow trout at dose of 25-50 mg/kg body cut of fish/day for 4-7 days. Other antibiotics like doxycyclin, kitasamycin, alkyl-trimtethyl-ammonium-calcium, oxytetracycline, josamycin, oleandomycin, lincomycin have also been used to control streptococcosis in cultured yellow tail in Japan. Reducing over crowding, overfeeding, unnecessary handling or transportation and the prompt removal and slaughter of all moribund fish in ponds or net cages at an early stage of infection may prevent outbreaks or reduce their severity.

a.3 Mycobacteriosis

Genus *Mycobacterium* consists of 54 recognized species of aerobic, non-motile, non-spore-forming, rod shaped bacteria which are characteristically, acid-alcohol-fast at some stage of growth. The species pathogenic for fish are *M. marinum, M. fortuitum* and *M. chelonae.*

Mycobacteriosis is piscine tuberculosis and affects a wide range of freshwater and marine species but particularly aquarium fish. Diagnosis is based solely on the identification of acid-fast bacilli in histological sections of tissue lesions. *M. fortuitum* and *M. marinum* affects fish from both tropical and temperature waters. Infections with *M. chelonae* has so far been identified only in cold water salmonid species of fish.

Piscine mycobacteriosis is a systemic chronic, progressive disease presenting various clinical features depending upon species and ecological conditions. The symptoms have been reported as lightlessness, anorexia, emaciation, dyspnoea, exophthalmia, skin discoloration and external lesions ranging from scale less to nodules, ulcers and fin necrosis as signs of advancing infection. In cold water salmonids there may be no external signs of disease other than mortality, or else variable degrees of skin coloration, stunted growth and retarded sexual development. Gross or microscopic grayish-white miliary granulomas may be found scattered or grouped in virtually any parenchymatous tissue but especially the spleen kidney and liver. The histopathology of naturally developing granulomas are characteristic of mycobacterial infections.

There is no suitable treatment for affected fish and diseased stock is normally destroyed and comprehensive disinfection of holding facilities should be under taken. Natural spread of infection within a closed environment may be controlled by the addition to the tank of chloramine B or T at 10 mg/litre for 24 h after which the water should be changed.

M. marinum, M. fortitum and *M. chelonae* are all capable of infecting warm blooded vertebrates, including man, *M. marinum* can cause cutaneous granulomas in man, usually of the elbow but also of the knee, fingers and feet resulting from abrasions incurred in swimming pools (swimming pool granuloma) or tropical fish aquaria.

a.4 Nocardiosis

The disease caused by *Nocardio asteroides* infections is called as Nocardiosis. Nocardiosis affects cultured yellow tail other tropical freshwater fishes. Trouts are also reported to be affected. Nocardiosis shares many clinical features with mycobacteriosis and is a chronic disease occuring sporadically.

Early signs of infection may include anorexia, inactivity, skin discoloration and emaciation. In later stages, modular caseous skin lesions which may ulcerate and internal distension due to the development of internal granulation may be observed. In yellow tail the appearance of cream colored modules on the gills has led to the description gill tuberculosis.

Transmission of infection could be directly from the natural environment. *Nocardia* can be readily isolated on standard bacteriology media like BHIA, TSA or NA. The only successful treatment reported is the oral administration of a combination of sulphisoxazole and doxycycline on minocycline. A clean environment is probably the most important factor in the prevention of disease.

3. Disease Caused by Anaerobic Bacteria

Only two species of anaerobic bacteria namely *Clostridium botulinum* and *Eubacterium tarantellas*, have been implicated as fish pathogens. *Clostridium botulinum* type E had been reported to cause botulism which is a chronic disease termed bankruptcy disease in farmed trout. *Eubacterium tarantellus* has been found to cause a neurological disease in striped mullet *Mugil cephalus*. The disease caused is called as eubacterial meningitis.

B. Bacterial Diseases of Shellfish

Commonly reported shrimp pathogenic bacteria are the members of the genus *Vibrio*. Other bacterial groups include *Aeromonas* sp., *Pseudomonas* sp., *Flavobacterium* and gram positive bacteria like *Mycobacterium*. Rickettsial and chlamydial diseases have also been reported from shrimps.

a.1 Vibriosis

Infections and disease caused by *Vibrio* spp. have been by far the most numerous of the reported bacterial agents of penaeid shrimp. *Vibrio* spp. were found to constitute the majority of cultivable bacteria associated with the gut, gills or cuticle of wild or cultured penaeid shrimp. *Vibrio* spp. are opportunistic pathogens which establish lethal infection as a result of other primary conditions that might include other infectious diseases, extreme environmental stress, wounds *etc.* Some recently occurring disease syndromes of penaeid shrimp have been caused by *Vibrio* spp. which behaves more like true pathogens than opportunistic invaders. Important Vibrio species causing infection are *V. alginolyticus, V. parahaemolyticus, V. harveyi, V. splendidus, V. vulnificus* and *V. damsela.* Bacterial infections in shrimp may manifest as (1) localized pits in the cuticle, which constitute the condition called bacterial shell disease (2) localized infections of the gut or hepatopancreas or localised infections from puncture wounds, limp loss *etc.* and (3) generalised septicaemia. All life stages of the penaeid shrimp may be affected.

Vibriosis in grow out ponds of penaeids have been reported from many countries. Major symptoms were cloudiness in hepatopancreas in PL, cloudiness of muscles in sixth abdominal segment, and progressive spots on gills and lymphoid organ. Histopathology revealed extensive necrosis and bacterial invasion of lymphoid organ, with multiple melanized hemocytic nodules. Nodules were found in all the organs while inflammation was found only in lymphoid organ. Affected shrimps lost their escape reflexes, had darkened colour and heavy fouling by epibionts. Less affected were pale and opaque, and gills were brown in colour. Specimens from the pond showed signs of body reddening, extended gill covers, slight melanized erosion of uropods, pleopods and periopods. They also had empty stomach and midgut except for a watery white liquid. Small black spots were present on the lymphoid organ. *V. alginolyticus V. parahaemolyticus* and *Pseudomonas* species were obtained from the haemolymph. Vibriosis occurred during 1989 and in Ecuador which was called Sindroma Gaviota (or Sea Gull Syndrome, SGS) where large number of sea birds were observed together on or near ponds and to feed on dead and dying shrimps. Hence, it was called as seagull syndrome.

Reduction of biomass by partial harvest or reducing stocking density and increasing the water exchange help to control vibriosis in culture ponds. For subsequent production cycles, it is best to dry to pond bottom until the bottom soil cracks. Excessive detritus have to be removed physically by desalting the pond bottom. Quick lime has to be applied at the rate of 0.5 kg/m^2 of pond bottom. Use of immunostimulants fed through the feed is found to provide better resistance to diseases by stimulating the non-specific defence mechanisms of shrimps. Use of probiotic bacteria is found to compete with *Vibrio* sp. there by decreasing the load of pathogenic organisms.

Vibriosis in Hatcheries

Luminescent bacterial infection in hatcheries cause larval moralities associated with luminescent *Vibrio* spp. infected larvae become weak and opaque. Larvae exhibits luminescence when observed in total darkness. Systemic infection results in mortalities in larvae and post larvae reaching up to 100 per cent of the population. The causative bacteria recorded from these penaeid hatcheries were strains of *Vibrio parahaemolyticus, V. alginolyticus, V. harveyi* and *V. splendidus.* The affected larvae refused to feed. Scanning electron microscopy (SEM) studies indicate that the vibrios colonise specifically the feeding appendages and oral cavity.

Sanitation and rigorous management practices help to control the infestation. The separation of mother shrimps and their faecal matter from the eggs has been done as soon as possible after spawning. Artemia nauplii being used as live feed should be rinsed before introducing into the hatchery during feeding. Initial bacterial load of the rearing water should be minimized by chlorinating and other forms of water treatment like ultraviolet irradiation and filtration. Thorough cleaning, disinfection using 200 ppm chlorinated water and drying of the larval rearing tanks after each cycle of operation helps to control bacterial load and eliminate bacterial pathogens.

Chemical control has limited use because of development of resistant strains of bacteria and limited tolerance of shrimp larvae to the drugs. Hence, strict water quality management and sanitation are recommended for preventing luminous vibriosis in penaeid hatcheries.

a.2 Bacterial Shell Disease

Bacterial shell disease lesions are typically brownish to black in colour. Single or multiple eroded areas on the general body cuticle, appendages, or gills may be present. Lesions may begin as relatively small local lesions (due to abrasions, puncture wounds, chemical trauma or other causes) that rapidly enlarge. This disease is also called as brown spot shell disease/burned spot disease/rust disease/ shell disease or black spot disease. *Vibrio, Pseudomonas* and *Beneckea* have been known to cause this disease. These organisms cause erosions and ulcerations of the cuticle by their ability to produce lipases, proteases, and chitinases. The black pigment in shell disease lesions is melanin, which is an end product of the crustacean inflammatory response. If not resolved by the shrimps responses, septicaemia and death will be the likely result. The infection can be controlled by provision of better water quality, removal of infected and dead prawns, reducing the stock and adequate nutrition help to control the disease. Reduce the organic load in the pond by increased water exchange. Minimise handling and avoid overcrowding and treatment with formalin 25 ppm in static condition for 24 hrs can be done.

Brown or Black Spot Disease

It is also called as shell disease, rust disease, burned spot disease, necrosis, *etc.* Infestation is caused by bacteria belonging to *Vibrio, Aeromonas* and *Pseudomonas.* Appearance of brownish to black erosion of the carapace, abdominal segments rostrum, tail, gills and appendages. In larval and post larval stages, the affected part shows a cigarette-butt like appearance. The infection is usually initiated at sites of punctures or injuries made from telson or rostrum, cracks on the abdominal segment from sudden flexture of the shrimp body, or from other damage caused by cannibalism. Progressive erosion of these exoskeletal lesions follows upon entry and multiplication of bacterial pathogens. The infection can be controlled by maintaining good water quality and keeping the organic load of the water at low levels by removing sediments, especially dead shrimps and moulted exoskeletons which harbour high numbers of bacteria on the lesions. Minimising handling and avoiding over crowding also helps. Infection can be treated with formalin 25 ppm in static condition for 24 h.

a.3 Filamentous Bacterial Infection

Growth stages affected by these bacteria are larvae, postlarvae, juveniles and adults. Signs are the presence of fine, colourless, thread- like growth on the body surface and gills. It can interfere with locomotory process and moulting can cause mortalities of PL with heavy infestations. In larger shrimps it can even result in respiratory distress. The bacteria involved in this infection are *Leucothrix* sp. Other types of filamentous and chain forming bacteria such as *Thiothrix, Flexibacter, Cytophaga*, and possibly *Flavobacterium,* alone or together with *Leucothrix* are also important agents of filamentous bacterial disease. The infection is treated by $KMNO_4$ at 5-10 ppm for 1hr in static treatment for 5-10 days.

a.4 Mycobacterial Infections

The causative agent of the infection is *Mycobacterium marinum.* Symptoms include abnormally dark pigmentation in areas of the body that hard multicolour melanized haemocytic nodules or larger prominent melanized granulomatous lesions composed of multiple nodules. Lesions due to this bacterium were observed in the lymphoid organ, heart, cuticle and systemically in the loose connective tissue of muscle, hepatopancreas, antennal gland, ovary and gills. No control measures or therapy for mycobacterium infection in penaeid shrimp have been reported so far.

a.5 Rickettsial and Chlamydial Diseases

These organisms are important pathogens of penaeid shrimp in wild and culture penaeid from diverse geographic locations. Chlamydial – like agent is observed in the cytoplasm of hepatopancreatic cells of infected *P. japonicus.* Heavy rickettsial infections of the hepatopancreas of the cultured shrimp have been found to induce reduced feed consumption and moderate mortalities. Rickettsial infections show gross signs of lethargy, inappetence, poor escape responses and a white coloured hepatopancreas. In case of systemic rickettsial infection in *P. monodon* reported from Malaysia and Indonesia, the rickettsia occurred within large basophilic, feulgen – positive cytoplasmic vacuoles where it formed microcolonies of 19 to 33μm in diameter. They were reported to occur along with other pathogens also.

From 1985 onwards a disease called as Texas pond mortality syndrome (TPMS) was reported from Texas due to Rickettsia – like agent in *P. vannamei* upto 50 per cent to 99 per cent stock were affected. Medicated feeds containing 1.5 to 2.0 kg of Oxytetracycline 1000 kg of feed has been found to control the infection. Because of their small size and inculturability by common microbiological methods, the rickettsia, chalmydia and related small bacterial forms get frequently over looked.

Chapter 3

Viral Diseases

Viruses are one of the smallest group of microorganisms with simple structure of a nucleic acid genome covered by a protein coat which are characteristically different from the rest of the prokaryotes in three aspects:

1. They have very small size ranging from 18 - 300 nm
2. They are metabolically inert particles, which means they need a living cell to carry out the biological function of replication
3. They have only one type of nucleic acid either RNA or DNA

Viruses are classified basically on the type of nucleic acid they contain as either DNA viruses or RNA viruses. Apart from the type of nucleic acid, the nature of the nucleic acid in terms of strandedness (single stranded or double stranded), segmentation (segmented or non-segmented genome), polarity of RNA genome (+ or –), linear or circular, molecular mass, number of open reading frames, presence or absence of a lipid bilayer envelope, symmetry of the capsomeres, size and density of virus particles, number of proteins and type of antigens and viral enzymes, geographic distribution and host range *etc.* are the other characters included in the classification of the viruses.

Viruses can be double stranded DNA or RNA viruses and single stranded DNA or RNA viruses. The genome of the viruses may be segmented or non-segmented, linear or circular. Based on the presence of an envelope they can be classified in to enveloped or unenveloped viruses. Shape of the viruses depend on the nature of arrangement of the viral nucleic acid and capsomeres. If arranged to form an icosahedron with 20 equilateral triangular surfaces, it is referred to as icosahedral symmetry. If the nature of arrangement is a helix, it is called helical symmetry.

Some of the viruses are also without any specific symmetry, which is referred to as complex symmetry.

Virus Cultivation

As the virus needs the help of a living cell for its multiplication, cultivation of the viruses could be undertaken only by specific methods. Traditionally, the viruses are cultivated by egg inoculation using hen's eggs. However this method has a disadvantage that all the viruses are not adaptable to eggs and thus may not be growing in them. Secondly they could be cultured in animals under bioassay system, where susceptible animals are inoculated with the viruses and the viruses are harvested from the tissues that support their multiplication. However, this raises a lot of ethical questions and not always practicable. The third and successful method is the use of cell culture systems, which ensure controlled and large scale multiplication of the viruses.

Virus Replication

Viruses replicate in the cell culture systems in variety of complex and biologically different methods. These however, follow general and well defined major steps. These include

Attachment

As a first step in the replication, the virus has to get attached to the cell surface. This is mediated by the presence of surface proteins of the virus particle and the specific receptors present on the cell surfaces. Viral surface proteins determine whether a virus could irreversible attach to the cell surface to initiate in infection.

Penetration

Once the virus is attached to the cell surface, it has to enter into the cell by crossing the plasma membrane. This is accomplished by endocytosis especially in the case of unenveloped viruses and by membrane fusion in enveloped viruses. Once in the cell, the virus is transported to the site of multiplication either cytoplasm or nucleus.

Uncoating

The virus genome has to be completely released from it capsid coat for the virus replication to begin. This process is known as uncoating.

Transcription

This is the process of generating mRNA from the genomic nucleic acid. Depending on the type of nucleic acid, the process would convert DNA or RNA to mRNA polarity.

Translation

The mRNA generated in the previous step acts as template for generating the viral protein in this step.

Replication of Nucleic Acid

Genome of the viruses are replicated based on the type of their nucleic acid.

Assembly and Release

The virions start to emerge when threshold levels of viral nucleic acid and protein molecule have accumulated in the infected cell. The proteins build up into nucleocapsids and later lead to packaging the genome into virus particles. Apart from self assembly in some cases, the assembly requires packaging signals to be associated with the required strand of the genome.

Release of the virus particles happens through different ways. Most of the enveloped viruses are released by budding from the cell membrane of the infected cells. Viruses also exit from the infected cells by lysis of the cell

Latency

In some of the infections, the viruses do not lead to productive infections but the virus goes to latency stage by either as a sequence integrated into the cell genome or as multiple copies of covalently closed circular DNA. The latency of infection of the virus can be activated at a later stage, a process known as induction. This could happen due to a variety of factors that by cells may be subjected to.

Effects of Virus in the Cell

Virus infection of a cell can lead to different outcomes.

1. Abortive infection – where the virus though infect the cell, gets eliminated due to the defence strategies of the cell
2. The infection lead to latent infection which can continue to be latent and stay as persistent infection or later lead to productive infection
3. Productive infection – the infection can lead to cell death and release of virus particles or become persistently infected with continuous release of virus particles without cell death
4. Apoptosis – the infected cell undergoes programmed cell death thereby preventing the spread of infection

Viral Pathogenesis

The complex series of events that involve virus replication and the host defence responses finally resulting in the induction of disease in the host are collectively known as viral pathogenesis. However, the replication of the virus at the primary site of multiplication need not always result in clinical disease.

The viral pathogenesis starts with the entry of virus in the host followed by primary replication and viral spread in the host. Viruses enter the host either through vertical transmission or by horizontal transmission. Viruses replicate at the primary site of entry and in fish it could be gills, gut or integument. Viruses spread within the host through blood or lymphatic circulation and based on the tissue tropism of the virus, further multiplication of the virus in the target tissue would result in the cellular injury. Host defence response would soon be initiated and depending on the effectiveness of the immune responses, the virus replication could be prevented and the cell injury could be minimised. An infection of the virus in the fish could lead to any of the following outcomes:

1. No clinical disease and elimination of the virus
2. No clinical disease but establishment of persistent infection (Carrier state)
3. Development of clinical disease and death of the fish
4. Development of clinical disease but recovery of the host and elimination of the virus
5. Development of clinical disease, recovery of the host but persistence of infection

A. Viral Disease of Finfish

a. DNA Viral Diseases

a.1 Koi Carp Herpesvirus (KHV)

Causes mass mortality in common carp and koi in many countries throughout the world. Based on its pathogenic effect in fish, it has also been termed carp interstitial nephritis and gill necrosis virus (CNGV). Infected fish exhibit lethargy, separate from the shoal and gather at the water inlet or sides of a pond and gasp at the surface of the water. They may also show hyperactivity loss of equilibrium and disorientation. Morphologically, pale discolouration or reddening of the skin, with rough texture, focal or total loss of epidermis, over - or under - production of mucus on the skin and gills, exophthalmia (sunken eyes) and haemorrhages on the skin and base of the fins and fin erosion are associated with this infection. There are currently no widely applied control methods for KHV. Artificially elevated water temperatures as a means to limit KHV infections and to induce anti-viral immunity are currently practiced for control.

a.2 Iridovirus

They have been associated with severe disease and economic loss in farmed food fish and ornamental fish, with mortality often reported to reach 50 per cent or more. Iridoviruses are reported from a number of ornamental fishes such as dwarf gourami *Colisa lalia,* orange chromide cichlid *Etroplus maculatus,* African lampeye *Aplocheilichthys normani* and other marine fishes. Diseased fishes display distinct histopathological signs of iridovirus infection such as systemic appearance of inclusion body-bearing cells, and necrosis of splenocytes and hematopoietic cells.

a.3 Epizootic Haematopoietic Necrosis (EHN)

It is caused by a double-stranded DNA, non-enveloped iridovirus, as causative agents of EHN. Moribund fish have loss of equilibrium, flared opercula and may be dark in colour. Fish may have enlargement of kidney, liver or spleen. There may be focal white to yellow lesions in liver corresponding to areas of necrosis. Acute focal, multifocal or locally extensive coagulative or liquefactive necrosis of liver, haematopoietic kidney and spleen are common. Necrotic lesions may also be seen in heart, pancreas, gastrointestinal tract, gill and pseudobranch. Basophilic intracytoplasmic inclusion bodies are noticed in necrotic areas in liver and kidney. The virus is found to infect perches and rainbow trout and it reported from Australia and Europe.

a.4 Lymphocystis Disease

Lymphocystis disease is caused by iridoviruses in a broad range of fish species. The disease is characterised by the development of macroscopically visible pearl-like or wart-like nodules primarily on body surface but also on internal organs. The infection is chronic, rarely fatal and in most cases is self limiting. However, the infection reduces consumer acceptance of the fish. The disease is diagnosed by the appearance of the nodules but can sometimes be confused with other infections such as "white spot" caused by *Ichthyophthirius multifilis*.

b. Diseases Caused by RNA Viruses in Fish

b.1 Infectious Pancreatic Necrosis (IPN)

The disease is caused by a highly contagious virus, infectious pancreatic necrosis virus (IPNV) belonging to the *Birnaviridae*. It affects both warmwater and coldwater fishes of freshwater and marine environment. IPNV enjoys a wide geographical distribution, reported from America, Europe and Asia. IPN characterized by sudden mortality with a progressive increase in severity. Cumulative mortalities may vary from less than 10 per cent to more than 90 per cent. The virus is also isolated from ornamental fishes in which it causes severe mortality. Affected fish die very quickly. In outbreaks fish loose appetite, becomelethargic and eventually disorientated. Some become ascitic and haemorrhages at the base of the fins are common. Corkscrewing/spiral swimming motion, darkening pigmentation, a pronounced distended abdomen, pop-eye and/or pale faecal casts are also seen in IPN infected fishes. In IPNV infection, survivors of the disease become carriers of the virus for the rest of the life. The IPN is transmitted both horizontally and vertically *via* the egg. Horizontal transmission is by viral uptake across the gills and by ingestion. The virus shows strong survival in open water conditions and can survive a wide range of environmental parameters. Prevention methods include avoidance of fertilised eggs from IPNV carrier broodstock, surface disinfection of eggs to reduce transmission rate, vaccination *etc.* Control of losses during outbreaks involves reducing stocking densities and dropping water temperatures (if possible).

b.2 Viral Nervous Necrosis (VNN)

This disease has been reported in at least 30 fish species, including freshwater ornamental fishes. The disease has typical characteristics of neurological abnormalities with clinical symptoms like abnormal swimming and whirling leading to mass mortalities of the young ones. The lesions of the disease develop in brain and retina. The common clinical signs noticed in the disease include abnormal swimming behaviour, pale colour, anorexia and swim bladder hyperinflation. Histologically, the larvae are characterised by vacuolation of cells of the nervous system, retinal layers and brain. Intracytoplasmic inclusions in brain cells are also noticed.

b.3 Spring Viraemia of Carp

This disease is one of the most important disease of cultured and ornamental fishes caused by spring viraemia of carp virus (SVCV), coming under the rhabdovirus group. This group of virus share many similarities with the swim bladder inflammation virus (SBIV) and both the viruses mostly indistinguishable. Clinically the infection is characterized by lethargy, darkening of the skin, exophthalmia, petechial haemorrhages of the skin and gills, inflamed vent, trailing mucoid faecal casts and loss of balance. Internal symptoms include viral septicaemia, oedema and necrosis in liver, pancreas, kidney, heart, brain, intestine and swimbladder, visceral haemorrhages and ascitic fluid in abdominal cavity. Swimbladder inflammation leads to the loss of balance and the fish often vertically hang in the water. The disease outbreaks are often associated with increasing water temperatures, and survivors become carriers of latent infection. During periods of stress, virus is shed with mucus and faeces and enters the uninfected fish through the gills that are considered the site of entry. As the SVCV is also transmitted by vectors, good husbandry practices are essential to keep the infection under check.

b.4 Viral Haemorrhagic Septicaemia

The disease is caused by a RNA virus belonging to the rhabdovirus genus and is related to the viruses causing infectious haematopoietic necrosis and spring viraemia of carp (SVC). Natural outbreaks of VHS occur in marine reared rainbow trout and turbot. Other fishes like sea bass and sea bream are susceptible to the virus and VHSV has been isolated from Atlantic herring, sprat, dab and Atlantic and Pacific cod in Alaska. In freshwater, rainbow and brown trout also get infected by VHSV. IN acute infection, clinical signs of the disease include darkening of body color, pop eye (exophthalmia), bleeding around the eyes, pectoral and pelvic fins, pale gills with pin point hemorrhage are found. Internally, petechial hemorrhages on the surfaces of fatty tissue, intestine, liver, swimbladder and in the muscle, presence of fluid in the abdominal cavity are the signs noticed. In subacute or chronic stage, some of the above signs are noticed with severe anaemia. Mortality is variable depending on the serotype ranging from 10 to 80 per cent. The virus is transmitted through the water, from infected carrier fish, birds, equipment, transport, water or by some blood sucking parasites.

b.5 Infectious Hematopoietic Necrosis (IHN)

The disease is caused by an RNA virus, which causes large scale mortality. During outbreaks, fish are typically lethargic with bouts of frenzied, abnormal activity such as spiral swimming and flashing. Affected fish exhibit darkening of the skin, pale gills, ascites, distended abdomen, exophthalmia and petechial haemorrhages internally and externally. A trailing faecal cast is observed in some species. Spinal deformities are present among some of the surviving fish. Internally, fish appear anaemic and lack food in the gut. Degenerative necrosis in haematopoietic tissues, kidney, spleen, liver, pancreas and digestive tract. The cellular debris, termed necrobiotic bodies, can be seen in stained tissue imprints from the anterior kidney and has diagnostic value.

b.6 Infectious Salmon Anaemia (ISA)

The disease is caused by an enveloped RNA virus, which buds from plasma membranes after replication. The disease is observed in Atlantic salmon. Experimentally, ISA has been transferred to sea trout and rainbow trout. Clinical signs of ISA are lethargy, haemorrhagic eyes, pale gills and a distended abdomen filled with fluid. The liver shows focal necrosis, petechial haemorrhage. Haemorrhage may occur in the stomach wall and the organ becomes filled with a serous or viscous fluid. The liver may be yellowish or pale with haemorrhage. Histological examination shows haemorrhagic, focal and confluent liver necrosis. Anaemia is accompanied by leucopenia. Mortality up to 90 per cent has been recorded. Only horizontal transmission has been recorded. Infected live salmon, and infected biological waste from slaughter and processing plants and water are the main source of infection. Sea lice may act as vectors. Towards control, ascertain management practices that reduce exposure to biological material. Avoid locations close to infected farms or close to slaughter and processing plants. Frequent removal of dead fish from net pens also reduces tile risk of ISA outbreaks.

b.7 Salmon Pancreas Disease (Pancreas Disease)

The virus responsible for the disease is identified as a togavirus and named salmon pancreas disease virus (SPDV). Pancreas disease (PD) is a subacute to chronic disease of farmed Atlantic salmon posts molts and growers. PD has been found in Europe and rainbow and brown trout are also susceptible. Clinical signs include reduced feeding response and slow movement near the water surface. As the disease progresses the fish stop feeding, become emaciated and frequently show a loss of balance. Absence of fat body, localised pinpoint haemorrhage and an empty gut are also noticed. Mortality 50 per cent is noticed. Pancreas disease is diagnosed by microscopical examination of stained tissue sections which shows areas of necrosis and fibrosis with total loss of the acinar cells of the exocrine pancreas. Pancreas disease has also been linked with the disease known as 'sleeping disease' (SD) in freshwater rainbow trout. PD has been experimentally transmitted by cohabitation and infective injection of kidney homogenate.

B. Viral Diseases of Shellfish

Currently, over 14 virus diseases of cultured shrimp are recognised. Each of these viruses is comprised of many strains and some of them are highly pathogenic to certain species while others are not harmful. The major viruses of shrimps are as follows:

Type	Family	Virus
ssRNA viruses	Dicistroviridae	Taura Syndrome Virus (TSV)
ssRNA viruses	Roniviridae	Yellow Head Virus (YHV)
ssRNA viruses	Roniviridae	Gill-Associated Virus (GAV)
ssRNA viruses	Bunyaviridae	Mourilyan virus (MoV)
ssRNA viruses	Totiviridae	Infectious Myo Necrosis Virus (IMNV)
ssRNA viruses	Rhabdoviridae	Rhabdovirus of Penaeid Shrimp (RPS)
ssRNA viruses	Togaviridae	Lymphoid organ vacuolisation virus (LOVV)
ssRNA viruses	Nodaviridae	Nodavirus/extra small virus *Macrobrachium rosenbergii* (MrNV/XSV)
ssRNA viruses	Nodaviridae	Laem-Singh Virus (LSNV)
dsRNA viruses	Birnaviridae	Infectious pancreatic necrosis-like virus (IPN)
dsRNA viruses	Reoviridae	Reo-like virus
ssDNA viruses	Parvoviridae	IHHNV Infectious hypodermal hematopoietic necrosis
ssDNA viruses	Parvoviridae	Hepatopancreatic parvovirus (HPV)
dsDNA viruses	Baculoviridae	Nonoccluded bacilliform virus or Baculoviral midgut gland necrosis virus (BMNV)
		PvNPV or (BP-type) Single-nucleocapsid polyedrosis PmN PV or(MBV-type)
		Single-nucleocapsid polyedrosis virus (MbNPV)
	Nimaviridae	White spot syndrome virus (WSSV)

a. Togaviruses

Toga - like viruses are observed in the lesions of lymphoid organs of many shrimps (*P. monodon, P. esculentus, P. vannamei, P. stylirostris,* and *P. chinensis*). Extreme hyperplasia and metastasis of the lymphoid organ (Oka organ) can be histologically recognised in the lesions. The virus particles of 30 nm diameter was observed in intracytoplasmic eosinophilic to pale basophilic cytoplasmic inclusion bodies occurring in cells with highly vacuolated cytoplasm. Cells were found to have pyknotic and karyorrhetic nuclei. The virus particles of 55 mm diameter are either by budding or by invagination and fusion of host cell membranes around the nucleocapsids

b. Yellow Head Virus (YHV)

YHV is an RNA virus reported only from *P. monodon* in Thailand. All ages of juveniles from 20 days post – stocking in the pond could be infected and mass mortalities up to 100 per cent within 3 - 5 days of first clinical signs are observed. The disease can be recognised by pale body colour with yellowish gills and hepatopancreas. It affects many tissues such as gills, lymphoid organ, hemocytes and connective tissue. Histologically, the infection can be diagnosed by the degenerative changes in nuclei and presence of cytoplasmic basophilic inclusion bodies.

c. Reo like Viruses

Reo-like viruses are reported from *P. japonicus, P. monodon, P. chinensis* and *P. vannamei*. Two separate types designated as type 3 and type 4 have been recognised among the isolates. Hepatopancreas has been suggested as the principal target for both the viral strains. Presumptive diagnosis of infections can be made with routine histology and staining. Feulgen - negative cytoplasmic inclusion bodies are demonstrated in I and R type cells of atrophied hepatopancreas. Transmission electron microscopic (TEM) pictures show crystalline array of virus particles of 50-60 nm diameter. Reovirus infections were always reported in mixed infections. Hence the role of reovirus as pathogens is not completely clear.

d. The Parvo-like Viruses

This group includes 3 viruses, HPV, LOPV and IHHNV. These parvo-like viruses are characterised by their small size (20 – 30 nm), single stranded DNA genome and unique cytopathology.

d.1 Hepatopancreatic Parvo-like Virus (HPV)

HPV has been widely distributed in Indo-pacific. In many cases HPV has been found to occur as dual infections with MBV. In cases of such infections, mortality rates can range from 50-100 per cent of the affected populations within 4 weeks of onset of the disease. HPV infections in penaeids in most cases are characterized by the presence of an atrophied and white hepatopancreas reduced growth rate, anorexia, poor preening activity, and occasionally occurring opacity of abdominal muscle followed by surface and gill fouling by epicommensals and secondary infections by opportunistic pathogens like *Vibrio* spp. In some cases of HPV infections, however, the animals are found to exhibit no outward signs of the disease.

HPV infections can be diagnosed by the histologic demonstration of single, prominent, basophilic, Feulgen-positive intranuclear inclusion bodies in hypertrophied nuclei of hepatopancreatic tubule epithelial cells. Compression and lateral displacement of nucleolus and margination of chromatin could also be visualised in HPV infected cells. In the early stages of development, HPV inclusions are small eosinophilic bodies centrally located in the nucleus associated with nucleolus. Cells in the distal portion of the tubules are most commonly affected. HPV has been reported from cultured or wild penaeids from Indopacific region including *P. chinensis* from China, *P. merguiensis* and *P. indicus* from Singapore, *P. penicillatus* in Taiwan, *P. esculentus* and *P. merguiensis* in Australia. *P. monodon* and *P. semisulcatus* from a number of locations in the Indopacific coast. In Indian shrimp farms, the HPV shows a low incidence rate.

d.2 Infectious Hypodermal and Hematopoietic Necrosis Virus (IHHNV)

IHHNV is a single stranded non - enveloped icosahedral DNA virus, smallest of the known penaeid shrimp virus with an average diameter of 22 nm. The virus has an average density of 1.40 g/ml, an estimated nucleic acid size of 4.1 kb and contains 4 virion polypeptides (74, 47, 39 and 37.5 kDa) making up the capsid. IHHNV is distributed worldwide with more prevalence in the Southeast Asia. *P. monodon* has been found to be the natural host of the virus.

IHHNV causes acute catastrophic epizootics in cultured juveniles of *P. stylirostris*. Severity of the disease is less in other penaeids, which are infected. IHHNV has been reported to be associated with runt - deformity syndrome in cultured *P. vannamei* Affected shrimp are found to exhibit reduced growth and a variety of cuticular deformities to rostrum, internal and other areas of exoskeleton. The infection is transmitted both horizontally and vertically. In *P. monodon*, the infection causes bluish coloration and opaque abdominal musculature. Histologically, prominent Cowdry type A eosinophilic intranuclear inclusions provides confirmative diagnosis of IHHN. IHHN results in hypertrophic nuclei of cells in tissues of ectodermal (epidermis, hypodermal epithelium of fore and hind gut, nerve cord and nerve ganglia) and mesodermal origin (hematopoietic organs, antennal gland, gonads, lymphoid organ, connective tissue and striated muscles). Survivors of IHHN epizootics carry the virus for life and transmit the virus horizontally and vertically.

d.3 Lymphoid Organ Parvo-like Virus (LOPV)

LOPV was found in cultured shrimps (*P. monodon, P. merguiensis* and *P. esculentus*) in Australia. Affected shrimps had multinucleated giant cells in their hypertrophied lymphoid organs. Giant cells showed nuclear hypertrophy, marginated chromatin and formed fibrocyte-encapsulated spherical structures. Giant cell nuclei had intranuclear inclusion bodies containing DNA. Electron microscopic studies revealed the presence of 25-30 mm diameter virus - like particles. There is a speculation that IHHNV and LOPV are the same agent.

e. Baculoviruses

Baculovirus in penaeids are grouped under (1) Type A occlusion body forming viruses BP and MBV and (2) Type C non - occluded baculoviruses BMN, TCBV, Owen's hemocyte–infecting baculovirus and WSDV. Although the baculoviruses are spread through horizontal transmission, some are demonstrated to transmit vertically. In hatcheries, BP and BMN often cause serious epizootics in the larval and postlarval stages of their hosts while MBV produce serious infections and mortalities in the late PL and juvenile stages of hosts.

e.1. BP Type Baculoviruses

BP (*Baculoviruspenaei*) causes significant disease in the larval, postlarval and early juvenile stages of many species of shrimps. The infections can be easily diagnosed by the demonstration of prominent tetrahedral occlusion bodies in unstained squash preparations of hepatopancreas, midgut or faeces and also in histological sections. In histological sections, occlusion bodies are found in single or multiple, eosinophilic, usually triangular within hypertrophied nuclei of hepatopancreatic tubule epithelial cells or in midgut epithelial cells. In hypertrophied nuclei, morphometric comparison of BP virions suggested the occurrence of 3 distinct strains of BP (Equator, Gulf of Mexico, Hawaii). BP is found to occur sporadically in hatchery infecting larval stages. BP has not been reported outside of the Americas and Hawaii

e.2. Monodon Type Baculoviruses

MBV enjoy a world-wide distribution, on the Indopacific coast of Asia, Antarctica, Africa and Southern Europe and have a diverse host range. Another viral agent called *Plebejus baculovirus* (PBV) found in cultured penaeids in Australia in reported to be probably a different strain of MBV. These are type A baculoviruses measuring 75 x 324.

MBV infections are diagnosed by the presence of single or multiple, generally spherical occlusion bodies in hepatopancreas and midgut epithelial cells. MBV occlusions in squash preparation can be easily visualized by 0.05 per cent aqueous malachite green staining, epifluorescence microscopy and acridine orange staining. The occlusion bodies contain randomly occluded enveloped virus aggregations of MBV virions especially near to the nuclear membrane. Virogenic stroma is often found at the centre of the nucleus while the nucleolus will be marginated to the inner surface of nuclear membrane and get modified to form virogenic stroma.

MBV was first discovered in a quarantined population of *P. monodon* that had originated from Taiwan. Currently, the virus is found distributed in Taiwan, Japan, Philippines, Indonesia and generally throughout Indopacific. Despite the world distribution of MBV, the virus is not apparently a highly virulent pathogen of *P. monodon*. MBV is found in healthy prawns and in disease epizootics, *P. monodon* has been found to frequently have mixed infections by MBV and other viral, bacterial or protozoan pathogens.

e.3. BMN (Baculoviral Midgut Gland Necrosis Virus) and other Type C Baculoviruses

This group does not produce occlusion body in the nuclei of infected cells. Infections by these viruses can be diagnosed by necrotic hepatopancreatic (midgut gland) tubule, epithelial cells that have hypertrophied nuclei with marginated chromatin, diminished nuclear chromatin, nucleolar dissociation and no occlusion bodies. Type C baculoviruses have been manually recognised in *P. japonicus* and *P. monodon*. BMN has been extensively studied. Enveloped viruses average 72-310 nm and may occur in dense aggregates especially near to nuclear membrane and surrounding virogenic areas.

BMN causes serious epizootics in hatchery reared *P. japonicus* in Southern Japan. The disease in characterised by sudden onset and high mortality rate. The disease appears from protozoea stage 2 and mysis larvae but is more severe up to PL

9 – PL 10. The disease subsides by PL 20. The infected larvae can be distinguished by its inactive floating on the surface and by a white turbid midgut line through the abdomen. Histological conformation of the infection could be demonstrated by the presence of necrotic hepatopancreatic tubule epithelial cells possessing hypertrophied nuclei with marginated and diminished chromatin, nucleolar dissociation and the absence occlusion bodies.

f. White Spot Syndrome Virus (WSSV)

WSSV, formerly known as systemic ectodermal and mesodermal baculovirus (SEMBV) is a non - occluded baculovirus – like agent infecting many penaeid species. The virus causes characteristic epizootic white spot disease of 2 - 7 days duration with mortalities ranging up to 80 - 100 per cent. The disease occurred in on - growing juvenile shrimp of all ages and sizes but mostly from one to three months after stocking in the grow-out ponds. WSSV outbreak occurs in all types of farming systems irrespective of stocking density, water quality and salinity. The infected shrimp swim to the surface of the water and gather at the pond dykes. Typical clinical signs include broken antennae, white spots of 1 mm size in the cuticle and/or reddish discoloration, empty guts and sometimes with cuticular epibiont fouling and lymphoid organ swelling. Histologically the disease was characterised by widespread and severe nuclear hypertrophy, chromatin margination, eosinophilic to large basophilic intranuclear inclusions with variable multifocal necrosis in most tissues of the animal. The virus is enveloped, rod shaped to elliptical and measured 292 x 111 mm. The white spot disease could be controlled by avoidance of infection and contamination.

g. Other Important RNA Viral Diseases

Other important viruses causing diseases in shrimps include taura syndrome virus, infectious myonecrosis virus, Laem-Singh virus *etc.*

Chapter 4

Parasitic Diseases of Fish

a. Classification of Parasites

Following are the important fish parasites belonging to different group of invertebrates.

Protozoa

Single celled eukaryotic organisms. Classification - Levine *et al.*, 1980

Phylum: Sarcomastigophora

Flagella/pseudopodia present and single nucleus.

Subphylum: Mastigophora

Class: Phytomastigophorea- contain chloroplasts in their cytoplasm -dinoflagellate parasites *Oodinium/Piscinoodinium* sp. and *Amyloodinium* sp.

Class: Zoomastigophorea - do not possess chloroplasts and have a varying number of flagella.

Order: Kinetoplastida - one or two flagella. *Cryptobia* and *Trypanosoma*.

Order: Retortamonodida- two to four flagella, bodinid parasites of *Ichthyobodo* sp.

Order: Diplomonadida - one to four flagella - *Hexamita* spp.

Subphylum: Sarcodina

Order: Amoebida - amoebae possess pseudopodia - divide by asexual fission.

Phylum: Apicomplexa

Class: Sporozoea.

Order: Eucoccidiorida. - *Haemogregarina, Eimeria, Cryptosporidia* and *Goussia*.

Phylum: Microspora

Obligatory and intracellular parasites.

Class: Microsporea

Microsporidium, Pleistophora and *Glugea*.

Phylum: Myxozoa

Class: Myxosporea.Order: Bivalvulida. Spore- 2valves. - *Myxidium, Sphaerospora, Ceratomyxa* and *Myxobolus.*

Order: Multivalvulida. Spores with 3 or more valves. *Kudoa.*

Phylum: Ciliophora

Class: Kinetofragminophorea. Oral ciliature slightly distinct from body ciliature.

Subclass: Vestibuliferia.Includes *Balantidium*.

Subclass: Hypostomatia. Includes *Chilodonella*

Subclass: Suctoria.Possess suctorial tentacles, adults are sessile. Includes *Trichophyra*.

Class: Oligohymenophorea. Oral ciliature distinct from somatic ciliature.

Subclass: Hymenostomatia. - *Ichthyophthirius*.

Subclass: Peritrichia – *Epistylis* and *Trichodina*.

Phylum: Platyhelminthes

Flatworms, dorso-ventrally flattened, bilaterally symmetrical and acoelomate lack an anus and specialized skeletal, circulatory and respiratory systems. Majority monoecious.

Class: Monogenea. The monogeneans - no intermediate hosts, - small worms< 3 cm in length. - haptor, armed with hooks or suckers.

Sub-classes Monopisthocotylea and Polyopisthocotylea.

Subclass: Monopisthocotylea. Gyrodactylidae (viviparous, parasitic on skin, gill and fins,0.3-1 mm) and Dactylogyridae (gill parasite, oviparous, 2mm).

Subclass: Polypisthocotylea.

Class: Digenea.-endoparasitic - metacercarial stages – encysted in fish

Class: Cestoidea - endoparasitic - scolex - parasitic in intestine

Phylum: Nematoda

Bilaterally symmetrical-cylindrical body tapering at both ends. They possess gut and are sexually dimorphic.

Phylum: Acanthocephala

Elongate cylindrical worms having retractile proboscis bearing hooks. They have no guts and the sexes are separate.

Phylum: Mollusca

The larvae of fresh-waterbivalve molluscs are often found attached to the gills and outer surfaces of fish. The larvae have thin bivalve shells often with little hooks on their inner edge.

Phylum : Arthropoda

Class: Crustacea - Crustaceans are bilaterally symmetrical animals with segmented bodies having jointed appendages. The body is covered with a rigid chitinoid exoskeleton.

Subclass: Branchiura. Body flattened dorsoventrally –prehensile suckers, They also have pre-oral proboscis. *e.g. Argulus*.

Subclass: Copepoda. Mainly ectoparasitic in fish. *Ergasilus*, *Lernaea* Cyclopidea, Caligidea and Lernaeopodidea.

Phylum: Annelida

These are segmented,coelomate worms with a muscular body wall. Important fish parasite under this group is leeches, which may also act as vectors for other pathogens.

Class: Hirudinea- They have asegmented body round or dorsoventrally flattened with anterior and posterior suckers. Leeches are ectoparasitic on fish.

Phylum: Chordata

The lampreys, or cyclostomes, which are eel-like, mostly fresh-water or anadromous, jawless fishes having around suctorial disc-like mouth with horny teeth attach to other animals and act as parasites.

Protozoa

Most protozoan fish parasites have a direct life cycle with no intermediate host.

Myxosporean parsites (*Myxobolus cerebralis*) and blood-parasitic flagellates and haemogregarines are also considered to have a life cycle with a secondary blood-feeding leech as an intermediate host.

Monogenea

Monogenean parasites have a direct life cycles without an intermediate host. Adult parasite lays eggs, which hatch to release free-swimming ciliated larvae known as oncomiracidia. These infect suitable host fish within a few hours. If they cannot find a host they die. The oncomiracidia, which are attached to a suitable host develop into adults. Gyrodactylids are viviparous and give birth to new individuals identical to the parent without any intermediate host.

Digenea

The fish parasitic digeneans are oviparous and the eggs laid by adults hatch to release a small ciliated free-swimming miracidium larva. These are alive for a few hours, by then if it could locate the first intermediate host of a gastropod or bivalve mollusc, it develops into free swimming cercariae after an asexual reproduction phase. These larvae survive for up to about 24 hours within which it finds a suitable second host. In some cases cercariae penetrate into a fish and mature directly to adult stage. Sometimes fish may act as an intermediate host and the cercariaen cysts within the fish to form the metacercarial stage. Metacercariae may survive for several years in fish and when another fish or any final host eats this it develops into adult.

Cestoda

Cestodes are oviparous and the eggs laid are passed in the faeces of the final host which hatch to release a free-swimming larva, the coracidium. This is eaten by an invertebrate copepod intermediate host. Coracidium develops into procercoid, a stage which can infect the fish. When a suitable secondary host eat the procercoid, it penetrates through the gut wall and encysts in the viscera or musculature where it develops to the plerocercoid stage. When this infected fish is eaten by a final host (fish, a bird or a mammal) the larvae completes the life cycle by becoming an adult.

Acanthocephala

The Acanthocephalans need one invertebrate host for completing their life cycle. They produce eggs which are passed through the faeces and then hatch into an acanthor larva in the intermediate host. The acanthor larva penetrates into host body cavity and develops into acystacanth which, when eaten by a suitable final host, develops into the mature parasite.

Nematoda

Eggs of oviparous nematodes or larvae of viviparous ones released by the host are ingested by an intermediate arthropod host. Further development of the nematode occurs in the intermediate host. The nematode larva penetrates through the gut wall into the viscera and musculature and get encysted. Once the infected fish is eaten by the final host - a fish, bird or mammal, the life cycle of nematodes is completed.

Mollusca

Larval stage of fresh-water bivalve molluscs is known as glochidia, which are obligatory parasites in fish. The glochidia larva released by the adult mollusc finds a suitable fish host and hold tightly on to the gills, fins or skin. These would later metamorphose to form a free living juvenile mollusc.

Copepoda

Eggs of parasitic copepods hatch to free- swimming nauplius larvae which moults to copepodid stage. There may be several copepodid stages before getting matured. Adult female copepods attach to the final host. The adult males are usually non-parasitic.

Leeches

Leeches have a direct life cycle. Cocoons laid by adults are attached to a substrate or held by parent hatch to produce young leeches.

Lampreys

Lampreys spawn in fresh-water and the larvae is known as ammocoete larvae, which are slender and worm-like. These may live for a number of years buried in mud and later metamorphose into the adult stage. The adults die after spawning.

b. Disease Caused by Parasites in Fish

a. Protozoal Parasites

The majority of the fish parasites which cause disease in fish include protozoal parasites. Typically, these parasites are present in large numbers either on the surface of the fish, within the gills, or both. When they are present in the gills, they cause problems with respiration and death will commonly occur when additional stressors are present in the aquatic environment. Protozoal parasites on the skin, fins or scales only, (*i.e.*, not affecting the gills) usually do not result in death, unless they are accompanied by a secondary bacterial infection. The more common protozoal parasites are listed below.

a.1 *Ichthyophthirius multifilis*

This is probably the most common parasite of all fishes. The common name for this parasite and disease is "Ich" or "white spot". The mature parasite reaches approximately 1 mm in diameter and is commonly observed in the gills and/or skin as coalescing white spots, hence the common name. The trophont or mature stage of the parasite has a large "horseshoe" shaped nucleus, and the entire surface of the parasite is covered in cilia. The life cycle of this parasite is direct, but is spent, in part, off of the host. The trophont is

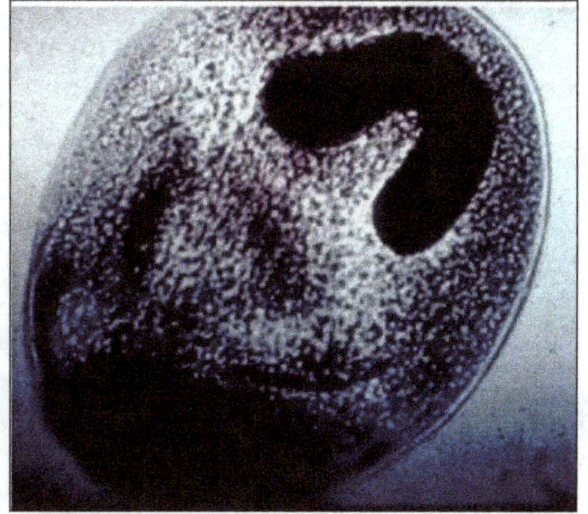

within the epidermis of the host, until it leaves the fish, encysts (demonstrates mature cyst of this parasite) and divides to produce many host-seeking tomites. The tomites penetrate the skin and gills of the fish to complete the life cycle. The life cycle is temperature dependent with a shorter life cycle occurring at warmer water temperatures.

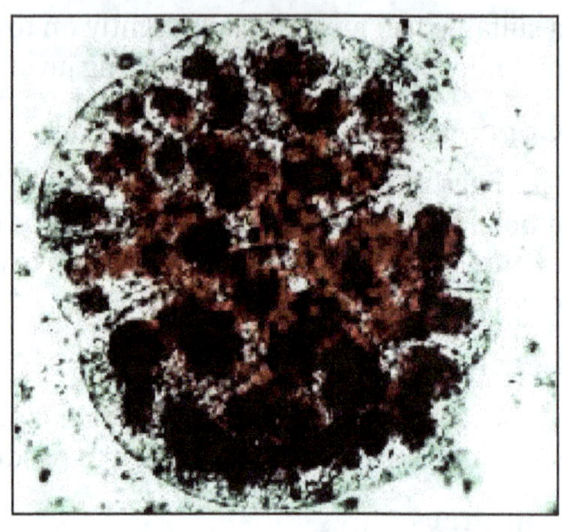

Fish with a cutaneous infection will "flash", *i.e.*, turn over and expose their white underside, whereas fish with a gill infection will "pipe", *i.e.*, come to the surface of the water and "breathe" through their mouth. Gill lesions include

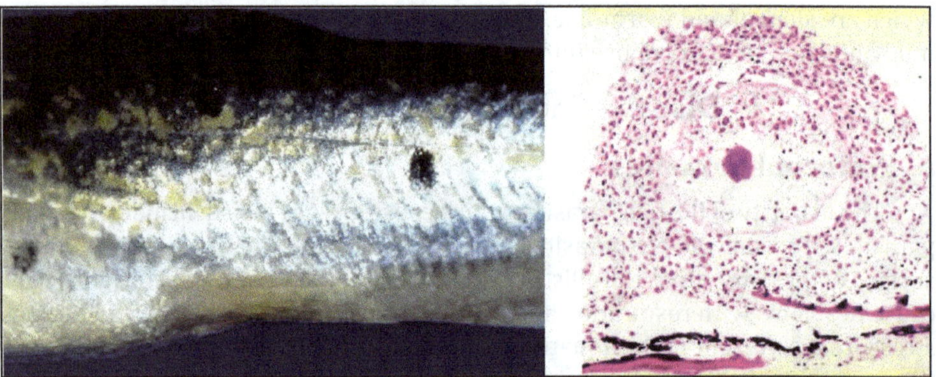

epithelial hyperplasia with the presence of mature trophonts within the gills. Cutaneous lesions also exhibit focal epidermal hyperplasia, with parasites being located beneath the hyperplastic epidermis. demonstrates a single "Ich" organism with hyperplastic epidermis.

a.2 *Trichodinia* sp.

There are three genera which form the Trichodina complex: Trichodina, Trichodonella and Tripartiella, however, all three are commonly referred to as "Trichodina". All are approximately 100 mm in diameter and have a saucer to "frisbee" shape and are ringed with cilia around its entire surface. They have a circular arrangement of tooth-like structures (denticular ring) within the body which provides them a characteristic appearance in fresh gill

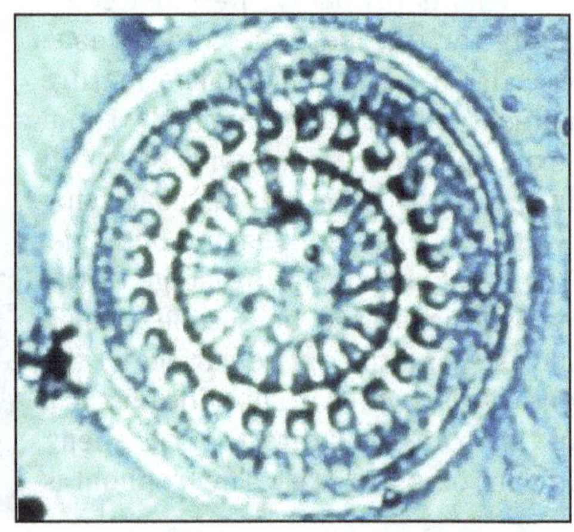

and skin cytology preparations. Fish with severe gill infections of trichodina will have respiratory and osmoregulatory difficulty and may "pipe" as well as "flash" if there is cutaneous involvement. Fin erosions and/or ulcerations can be observed in chronic cutaneous infections. Diagnosis of this parasitic disease is dependent upon identification of the parasite within the skin or gill cytologic preparations or histopathology.

a.3 *Ambiphyra* sp.

These are ciliated protozoal organisms which are thought to be free-living, but have been known to parasitize fish. They are sessile organisms with a cylindrical to conical body with oral cilia and a permanent motionless equatorial ciliary fringe. They range in size from approximately 60-100 mm and adhere to the epithelium of the skin and/or gills. Disease and death of fish have been associated with chronic infections of the gills due to mechanical blockage of respiratory epithelium. Diagnosis of this parasite is dependent upon identification of this organism within the skin or gill scrapings or histopathology.

a.4 *Ichthyobodo* sp.

This parasite is also known as Costia sp. These are obligate flagellate parasites with a direct life cycle. The free-swimming form is renniform, and approximately 10-20 mm long with two pairs of flagellae; whereas the attached form is pear-shaped and attaches to the gills and skin. Disease associated with this parasite includes increased cutaneous mucous production (hence the lay terminology of "blue slime disease"), epithelial hyperplasia of the skin and gills, ulceration and erosion of fins. This pathogen commonly causes disease in salmonid fry, resulting in high mortality. Diagnosis is dependent upon demonstration of the agent within affected fish by cytology or histopathology.

a.5 *Hexamita* sp.

These parasites are pyriform to oval with tapering toward the posterior end. Occasionally, rounded individuals can be identified. The organisms are 6-8 mm wide and 10-12 mm long. They have three pairs of anterior flagella which are approximately one and one-half times the length of the body. The flagella originate from the blepharoplast at the anterior end of the axostyles. These organisms can reproduce by longitudinal binary fission as well as undergo schizogony within the epithelial cells of the ceca or intestine. This parasite causes disease within the gastrointestinal tract of fish and affected fish will have clinical signs related to malnutrition and emaciation. Diagnosis is dependent upon finding the parasite from cytologic scrapings of the ceca or intestinal tract or histopathology of these organs. A poorly understood parasite, thought to be Hexamita-like is thought to be responsible for Freshwater hole in the head and lateral line erosion (FHLLE).

a.6 *Chilodonella* sp.

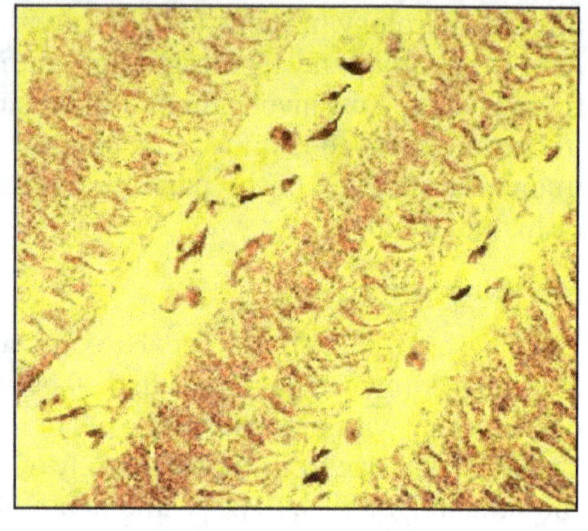

This is a motile ciliated protozoal parasite which causes disease in the skin and gills of fish. It is typically heart-shaped with the posterior end being broader and slightly notched. It measures approximately 20-40 mm in width and 30-70 mm in length and its surface is covered with cilia. There is a large macronucleus in the posterior portion of this organism and a smaller micronucleus is near or within the macronucleus. This parasite has been attributed to death of fish due to respiratory and osmoregulatory imbalances associated with severe gill parasitism. Diagnosis is dependent upon demonstration of the organism within the affected organs by either cytology or histopathology.

b. Myxozoan Parasites

This is a very large group of parasites which can cause disease in a wide variety of fishes. They are obligate parasites of tissue (histozoic forms that reside in intercellular spaces or blood vessels that reside intracellularly) and organ cavities. Key characteristics of the Myxozoa include development of a multicellular spore, presence of polar capsules in their spores and endogenous cell cleavage in both the trophozoite and sporogony stages. The method of transmission of myxozoans is unknown, but evidence suggests that at least some pathogenic myxozoans have an indirect life cycle. This life cycle may require the completion of two different life cycles involving a vertebrate (fish) and an invertebrate (annelid) host with each life cycle having its own sexual and asexual stages. Severe infestations by these parasites can result in disease and/or death of the host fish. Each parasite is somewhat species specific as well as organ specific. A few of the more common myxozoan parasites are discussed below.

b.1 *Myxobolus cerebralis*

This parasite is known for causing "whirling disease" in salmonids. This is a chronic debilitating disease which is very common in the Western U. S. and is uncommon in the Midwestern and Eastern U. S., although recent "outbreaks" have been reported in New York. The parasite feeds on cartilage of the axial skeleton and clinical signs are related to this damage. This parasite produces spores within the cartilage

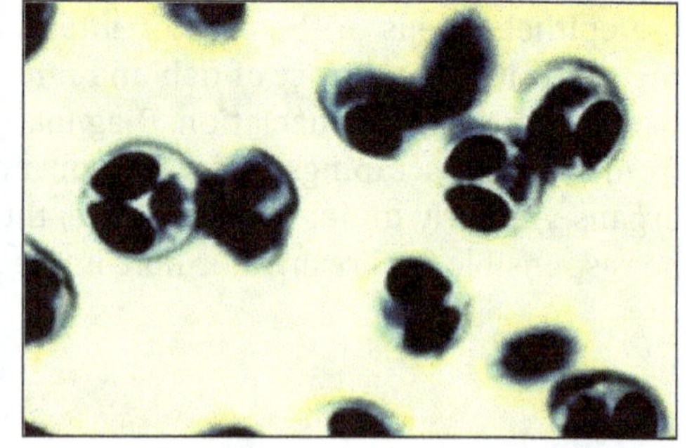

which are oval to circular and approximately 7-9 mm in diameter and 6-7 mm long with a thick mucoid envelope on the posterior half of the spore (demonstrates spores recovered from a fish with whirling disease). Histologically, myxozoan parasites within the cartilage of the axial skeleton, with morphological characteristics of *Myxobolus cerebralis*, must be observed to confirm the diagnosis.

b.2 *Aurantiactinomyxon*

This is currently thought to be the causative agent of "proliferative gill disease" of catfish. This myxozoan parasite causes a rapid, severe, proliferation of the gill epithelium which results in impairment of respiration and osmoregulatory function. The intermediate host is thought to be a microscopic aquatic oligochaete worm, Dero digitata, which is found in the mud and sediment in the ponds of affected catfish. Diagnosis is dependent upon the presence of the parasite within swollen, clubbed proliferative gill lamellae, many of which are fractured due to the associated chondrodysplasia.

b.3 *Henneguya* sp.

These parasites were once thought to be responsible for the "proliferative gill disease" of catfish, but are now thought to be less pathogenic, although they can cause respiratory problems if present in large numbers. They have a very characteristic paired, long, whip-like caudal processes giving them a spermatozoan-like appearance. Size of spores range from 8-24 mm in length with the caudal processes being approximately 20-45 mm long. Diagnosis is dependent upon observing the parasites within the histopathology of the gill lamellae (photomicrographs and demonstrate mature Henneguya cysts within the gill lamellae).

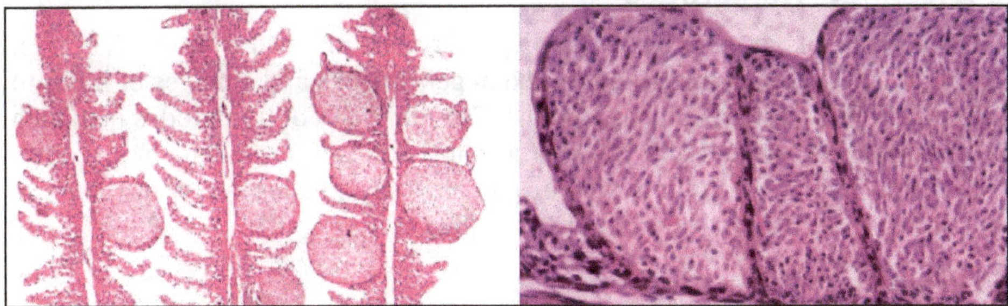

The phylum Platyhelminthes is composed of three taxonomic classes: Turbellaria, Trematoda, and Cestoda. The Turbellaria are all free-living and have no association with fish diseases. All members of the other two classes live in close relationship with host animals. The trematodes are commonly known as flukes, whereas the cestodes are known as tapeworms.

c. Mongenetic Trematodes

This is a group of trematodes which complete their entire life cycle on the host. The adults attach to the host by a haptor or opishaptor which is a specially adapted structure on the posterior end of the parasite. This organ has hooks which allow the parasite to attach firmly to the host fish. These parasites usually cause minimal damage to fish, but will infest the skin, fin and gills of pond fishes. Severe infestations may be responsible for poor respiration and/or emaciation. The two most common monogenetic trematodes include: Dactylogyrus and Gyrodactylus.

c.1 *Dactylogyrus* sp.

This parasite is approximately 0.2 to 0.5 in length, reaching a maximum length of 2.0 mm. It has seven pairs of marginal hooks and usually one pair of median hooks on the opishaptor. The dactylogrids have two to four pigmented spots (known as "eyes" or "eye spots") in the anterior fourth of the body. All dactylogrids are oviparous with no uterus.

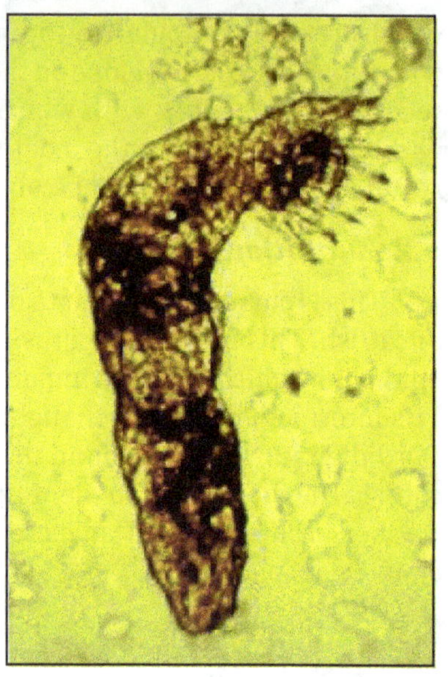

c.2 *Gyrodactylus* sp.

This parasite is smaller, rarely reaching a maximum length of over 0.4 mm. All species are viviparous with one to three daughter generations being observable in the V-shaped uterus. This parasite is more commonly found on the skin and less commonly present in the gills, although severe infestation will have both organs affected.

d. Digenetic Trematodes

This group of parasites has a complex life cycle with several successive larval generations, alternating sexual and asexual generations and changes of hosts to develop into the adult in its primary host. The life cycles of trematodes involving fishes may either use fishes as the primary hosts or as intermediate hosts. Adult trematodes may infest the intestine or gall bladder of fishes. A few of the more common digenetic trematodes are listed below.

d.1 *Diplostomum spathaceum*

The life cycle of this parasite begins as an adult trematode in the intestine of gulls or other fish-eating birds. The body of the adult is 0.3-0.5 cm in length and distinctly divided into a flattened anterior forebody and a cylindrical and narrower hindbody. Eggs are shed and passed in the feces of the bird to the water. The eggs hatch in approximately 21 days into free-swimming ciliated miracidia. The miracidia infest aquatic snails as the first intermediate host by penetration of the snail's hepatopancreas. The miracidia then become a mother sporocyst, followed by one or more daughter sporocysts. Each daughter sporocyst produces many cercariae which are released into the water. These cercariae seek a second intermediate host by penetrating the fins, skin, gills or cornea of small fishes. Primary host fish which ingest the initially infected fish (second intermediate host) become infected and the life-cycle is completed when the host fish are ingested by fish-eating birds.

d.2 *Posthodiplostomum minimum*

This trematode has several synonyms including: *Neodiplostomum minimum, Neodiplostomum orchilongum* and *Postodiplostomum orchilongum.* The life cycle of this trematode is very similar to that of *D. spathaceum* above, although, infectivity

of cercariae to fishes lasts no more than 24 hours after release from the snail. Each cercaria actively raises a scale and enters under the scale pocket, causing irritation to the fish. Blood, congestion and hemorrhage occur at the bases of fins or other places of cercarial penetration. The trematodes migrate from the point of entry to visceral organs of the fishes, usually within one to three hours after penetration. Metacercariae are located in any organ of the fishes' body, but are generally more numerous in the liver, kidney, heart, spleen and other organs of abdominal viscera. With many of the digenetic trematodes, the metacercariae within the skin results in increased melanin deposition, hence the term "black spot disease". (demonstrates "black grub" in the fin of an affected fish). Visible white or yellow spots in the visceral organs, usually no larger than 1 mm in diameter are often referred to as "white grubs" or "yellow grubs" (demonstrates "yellow grub" in muscle tissue of affected fish) and could be caused by several trematode species. Diagnosis of

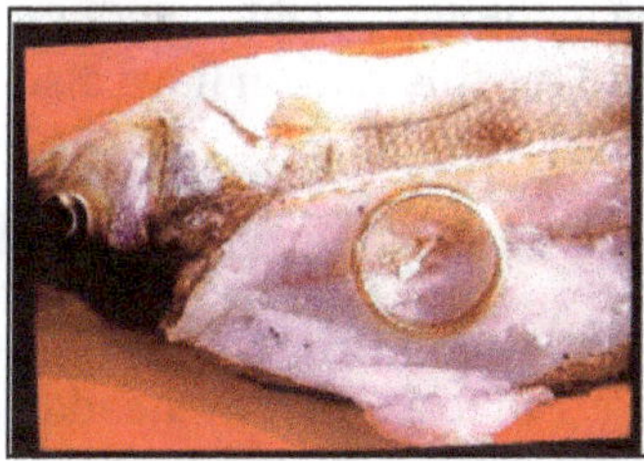

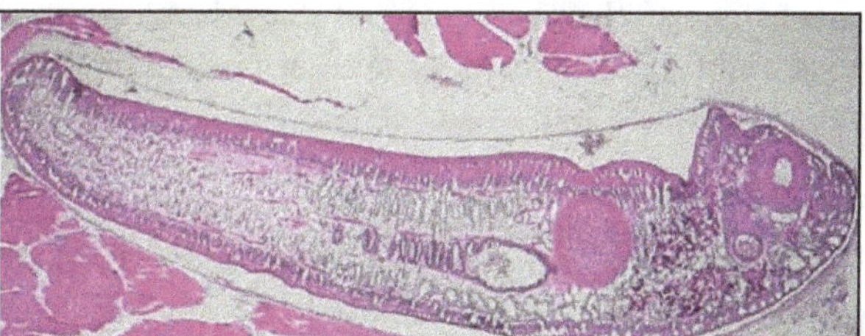

digenetic trematode infections is dependent upon identification of the genus and species of the trematode within infected fish.

e. Cestodes

Cestodes are a taxonomic class of organisms in which the adult stage usually lives in the intestinal tract of vertebrates. Intermediate stages live in a wide variety of body locations in both vertebrate and invertebrate hosts. The bodies of most cestodes are ribbon-shaped and divided into short segments called proglottids, hence the name "tapeworm". Diagnosis of cestodiasis is dependent upon 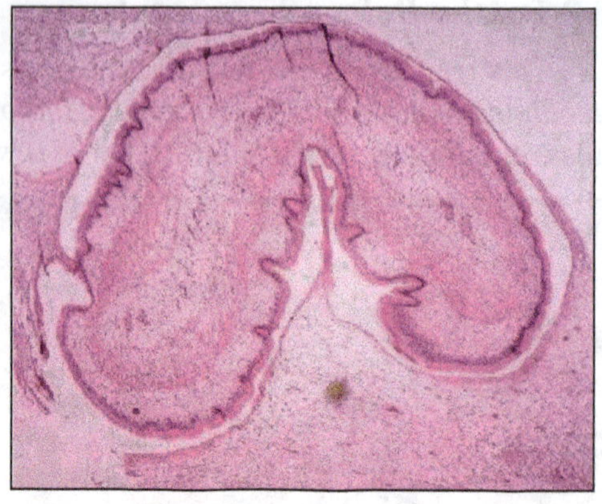 demonstration of the parasite within the intestinal tract of the fish. Clinical signs of cestodiasis include emaciation, anemia, discoloration of the skin, and susceptibility to secondary infections. Low numbers of pleurocercoids may be located in vital organs such as the brain, heart, spleen, kidney, or gonad and have a devastating effect on the fish (demonstrates a larval cestode within the liver of a fish). A few of the more common cestodes are listed below.

e.1 *Proteocephalus ambloplitis*

This parasite belongs to a large family of cestodes with ten recognized subfamilies. This parasite has a complicated life cycle involving a piscine second intermediate host and a piscivorous primary host fish. These parasites are observed within the intestines of bass. Eggs are expelled from gravid proglottids and pass from the host in feces. The eggs mature to an embryo which is ingested by several species of copepods as the first intermediate host. The procercoid develops from the embryo inside the copepods. The copepod is ingested by a forage-fish whereby the procercoid penetrates the intestinal wall of the second intermediate host. Some encyst in the wall of the intestine, other penetrate organs in the visceral cavity and may eventually reach the musculature. Here they develop into plerocercoids, which is then ingested by a piscivorous fish.

Ligula intestinalis: This group of parasites is distinct for three reasons: (a) they are not highly host-specific, but can develop in a wide variety of second intermediate host fishes, (b) the pleurocercoid stage develops sexually in the second intermediate host, and (c) these cestodes are very broad in shape and for this reason have also been known as "beltworms".

e.2 Acanthocephala

This group of parasites is comprised of worms with an anterior proboscis covered with many hooks. These parasites are often referred to as "thorny-headed worms". The body is composed of a presoma (proboscis and associated structures)

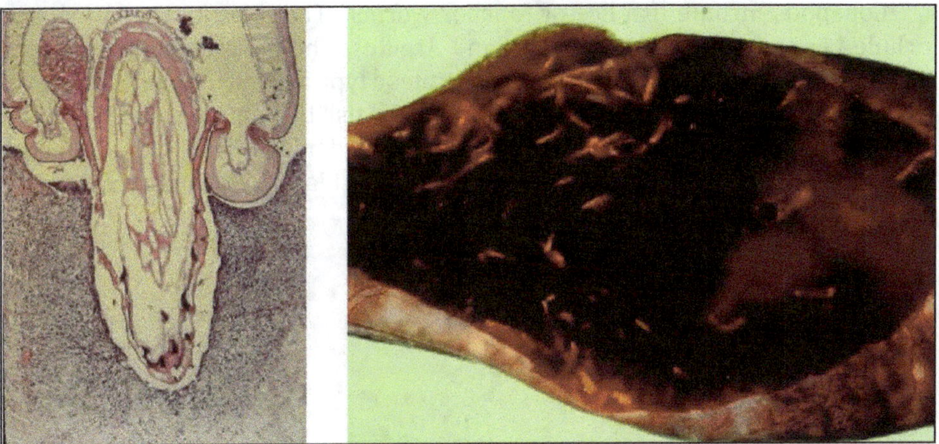

and a cylindrical trunk. The intestinal tract of the affected fish may contain a blood-tinged fluid, and histologically, the proboscis of these parasites will be firmly attached to the intestinal mucosa. demonstrates the proboscis (top) of an acanthocephala parasite which has burrowed into the intestinal mucosa (bottom) of an affected fish. Affected fish will exhibit emaciation, lethargy, anemia, and possibly death with a marked infection. We have observed salmonids from the Great Lakes with marked infestation of this parasite with literally hundreds of these parasites embedded in the intestinal mucosa.

f. Leeches and Copepods

Although not a common problem, occasionally, fish will be observed infected with either leeches or copepods. Leeches have long, slender flexible bodies and actively swim for an attack on their prey. Skin and underlying soft tissues are damaged and allow blood to flow into the leeches digestive tract. Leeches are not host-specific, and the damage to the skin and gills is dependent upon the number of leeches present at any time. Small fishes can be seriously injured or die due to excessive leech infestation.

Copepods include fish lice or "anchor worms". The more common fish lice include Lepeophtheirus and Caligus, and Argulus. The most common genus of anchor worms includes *Lernae* sp. (demonstrates a typical *Lernae* sp. anchor worm). All of these are external parasites which affect the fish by imbibing blood from the host fish and causing localized skin and soft tissue damage. They may also allow for secondary bacterial infection of the skin or musculature which may ultimately cause the demise of the fish.

Chapter 5

Fungal Diseases of Fish

A. Fungal Diseases of Finfish

Primary fungal infections of fish are considered to be rather rare, however, we have made this diagnosis in fish on more than one occasion. Nevertheless, by far, the most common fungal infections of freshwater fish are due to a group of fungi referred to as "water molds". More correctly, these are fungi belonging to the class Oomycetes, and most are due to the family Saprolegniaceae which is discussed below (Saprolegniasis).

a. Cotton Wool Disease

It is fungal disease of fishes and fish eggs caused by member of family saprolegiaceal. Characterized by presence of cotton like, white to gray growth on the skin, gills or eyes of fishes or in fish eggs.

Several investigators considered the saprolegniasis as the most significant mycotic disease affecting wild and culture fish causing economical losses, because the disease is difficult to prevent and treat, particularly in the intensive fresh water system, and is reported to be second only to bacterial disease.

Outbreaks of *Saprolegnia* spp. in farmed fish are usually restricted to chronic but steady losses. However, egg mortality can increase rapidly, causing significant numbers to die that may have a major economic impact. During severe winter, some farmers in the USA have reported losses of up to 50 per cent in farmed catfish, with an annual economic cost of about 40 million dollar. In Japan, the infection of coho salmon with *Saprolegnia parasitica* caused significant losses exceeding 50 per cent per annum. In Egypt, the disease reported in wild fish and culture fish causes mortality of fish and fish eggs.

Causative Agent

Saprolegnia is the main genus of water molds responsible for significant fungal infections of freshwater fish and fish eggs. *Saprolegnia parasitica, Saprolegnia diclina* and *Achlya hoferi* are the major etiological agents of saprolegniasis. A broad range of media has been used for the culture of fungi from fish. Often, the media are supplemented with antibiotics and a low-nutrient medium is preferred to reduce growth of saprophytic species and bacteria. The media are glucose yeast extract agar, glucose peptone agar, Sabourauds dextrose agar and corn-meal agar.

The fungus forms long, branched and non septated hyphae. Incubation temperature ranges between 5 and 37°C, however, temperature of 10, 15 and 20°C are the most common.

The hemp seeds also is used for isolation and identification the saprolegnia species. Saprolegnia has complex life cycle, which includes both sexual and asexual reproduction.

1. Sexual reproduction involves the production of the antheridium and oogonium.

2. Asexual spores of Saprolegnia release motile, primary zoospores. These are active only for a few minutes before encysted, germinated and release a secondary zoospore. The are motile for long period of time, they are considered the infectious spore of Saprolegnia.

Fish have three types of defenses against Saprolegnia:

1. Physical renewal of attaches spores by the renewal of mucous.

2. Morphogen in the mucous inhibited the growth of mycellium but not killing it.

3. Cellular response in the mucous have direct effect at growing mycellium.

Predisposing Factors

1. Malnutrition among cultured fish.

2. Presence of toxic substances in food or water or damage to skin, fin or gills from rough handling or external parasites may lead to secondary invasion by *Saprolegnia* spp.

3. Physical stresses such as reduced water temperature, high or low pH or high salinity may be responsible for invasion of fish by Saprolegnia.

4. Over crowding in fish culture increase the incidence of saprolegina infection.

5. High organic load is also identified as a cause of increase infection by *Saprolegnia* spp.

6. Saprolegeniasis may be secondary to other infection *e.g.* fin rot, ulcer disease and other bacterial disease.

Susceptible Species

Most species of fresh water fish are susceptible to take the infection with Saprolegnia, especially those fish in intensive aquaculture. Saprolegnia also infects the fish eggs by adhesion and penetration of egg membrane and can spread from dead eggs to healthy one.

Mode of Transmission

The disease is transmitted by:

1. Direct contact between diseased fish or fish eggs and healthy one.
2. Indirect contact through several sources, including, the water supply, transport vehicles, movement of staff between aquaculture facilities and farm equipment, such as nets.

Clinical Signs

Saprolegniasis is characterized by:

1. The appearance of cotton -like, white to grey growth on the skin, gills, fins and eyes or eggs of fish. These growth is white or grey white in colour.
2. In severe cases, 80 per cent of body may be covered with fungal growth.
3. In early infections, skin lesions are gray or white in color with a characteristic circular or crescent shape, which can develop rapidly causing destruction of the epidermis.
4. Lethargy of fish and loss of equilibrium.
5. Scales are lifted away from body surface of fish.
6. Neccrosis of fins and membranous part of gills may occur.
7. Respiratory manifestations appear on fish when infection is associated with gills.
8. Infected eggs are opaque in color with growth of fungus on eggs surface lead suffacation and become good medium for growth of the fungi. Saprolegnia lead to death by heamodilution or osmoregulatory failure.

Time to Death by Saprolegniasis Depends on

1. The initial site of infection.
2. Type of tissuse destroyed.
3. Growth rate of fungus.
4. The ability of the fish to withstand the stress of a fungus invasion.

Diagnosis

1. Case history.

2. Observation of a cottony, proliferative growth on the skin or gills alerts the clinician to a possible diagnosis of saprolegniasis.

3. Direct smear from fungal growth, presence of long, branched non-septate hyphae help in diagnosis of saprolegniasis.

4. Isolation and identification of saprolegnia using cultural method.

5. Recently: Polyclonal and monoclonal antibodies raised against various species of saprolegnia indicate that developing a rapid antibody assay for the detection of saprolegnia infection of fish.

Treatment and Control

Several investigators treat the saprolegnia infection of fish and fish eggs with formalin, copper sulfphate, potassium permanganate, sodium chloride, malachite green, acetic acid, hydrogen peroxide and iodophors using different doses at different intervals depending on fish species, severity of infection and climatic conditions. Water moulds can not be eliminated from any culture system. For this reason, prophylaxis is the best strategy for prevention and control of the saprolegnia infection, which is summarized in:

1. Removal the predisposing factors.

2. Avoiding damage of skin during transportation of fish.

3. Right kind of food with sufficient amount must be provided to fish.

4. Over crowding of fish must be prevented.

5. Preventing the introduction of new fish to the fish farm until known that fish are free from disease.

6. Disinfection of the equipments and utensils to prevent spread of the infection.

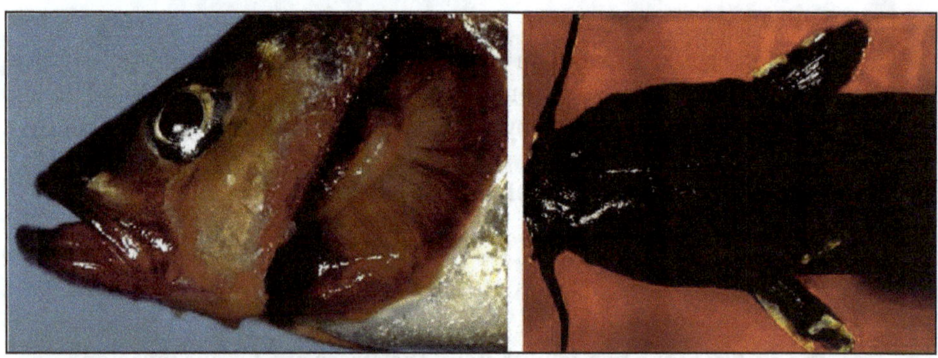

b. Branchiomycosis (Gill Rot)

Definition

It is a fungal disease involving gill tissues, affecting the most species of freshwater fish. The disease is caused by *Branchiomyces sanguinis* and

Branchiomyces demigrans. It is characterized by areas of infarctive necrosis of the gills, anorexia, and marpling appearance of the gills.

Causative Agent

There are two species from genus Branchiomyes, which are responsible for occurrence of the disease.

1. *B. sanguinis:* It grows mainly in the blood vessels of gill arches, filaments and in the gill lamellae.
2. *B. demigrans:* This fungal species is found in the parenchymal tissues of the gills.

Both species produce branched and non-septate hyphae. The fungi grow at temperature between 14 and 35°C. The optimun growth temperature appears to be between 25 and 32°C. The fungi grow on Sabouraud's dextrose agar medium.

Stress Factors

1. Elevation of the water temperature above 20°C.
2. Low dissolved oxygen.
3. Reduced water flow.
4. Over crowded conditions.
5. High levels of nutrients in the water and phytoplankton blooms.

Susceptible Species

The most species of freshwater fishes are susceptible to branchiomycosis.

Mode of Transmission

Fungal spores are transmitted by water to gills. These spores adhere to the gills, germinate and produce hyphae. The hyphae penetrate gills epithelium or within the blood vessels of gills depending on species of fungi.

Incubation Period

Incubation period for disease is related to water temperature. The disease has been rapidly developed within 2-4 days under suitable conditions.

Clinical Signs

1. Fish become weak in movement.
2. Fish are gathered in groups at water inlet and die.
3. Fish do not react to the approach of man and can be caught by hand.
4. There are respiratory distress in infected fish and do not swallow the air.
5. Fungus develops on or in gill tissue, or penetrates the blood vessels causing obstruction, congestion and necrosis of gill tissues.
6. Gills may be appearing red from impaired circulation.

7. Subacute form of branchiomycosis characterized by marbling appearance of the gills due to pale anemic patches in contact with red congested one due to disturbance of circulation in gills.

Diagnosis

1. Case history.
2. Clinical signs.
3. Microscopical examination of wet preparation from infected gill.
4. Isolation and identification of the causative agent.

Prognosis

Morbidity rate among fish populations with epizootics of brachiomycosis usually reach 100 per cent depending on fish species and susceptibility. Mortality rate may each 30 to 50 per cent of the fish population during late summer epizootics.

Treatment and Control

1. Strict sanitation and disinfection are essential for disease control.
2. Dead fishes should be collected and daily and burned or deeply buried.
3. Ponds with enzootic branchiomycosis should be dried and treated with calcium oxide (quicklime) or 2 to 3 kg copper sulphate per hectare.
4. Diseased fish can be treated with malachite green at 0.1mg/l for extended periods of time or 0.3mg/l for 12 hours.
5. Transportation of infected fish areas to non-infected areas must be prevents.
6. Increase of water supply help in control of that disease.
7. Stress factors must be avoided.
8. Regulating the feeding rate during warm weather.

c. Visceral Mycosis due to *Sporobolomyces salmonicolor*

This is an uncommon disease of salmon, which we have reported from this lab. This case involved high morbidity, low mortality with emaciation, cutaneous discoloration and ascites in chinook salmon fry. Lesions observed histologically included: aerocystitis (inflammation of the swim bladder), myositis, peritonitis, and dermatitis. The diagnosis was made by fungal culture of *Sporobolomyces salmonicolor*, (so named for its characteristic salmon color in culture) and identification of morphologically consistent fungal hyphae within tissue sections of affected fry.

d. *Ichyophorus hoferi*

The diseases caused by *Ichthyophorus hoferi* a fungus that causes an internal infection and is generally chronic and progressive. It usually is detected on necropsy when the characteristic spherical cyst stages are seen microscopically in the smears of granulomatous lesions of the heart, liver, spleen, kidney, skin, and muscle. Preventive measures include removing infected fish and avoiding feeding raw fish products. Iodophors of varying iodine concentrations are used to prevent mycotic infections of nonfood-fish eggs, which can be disinfected by a 100 ppm iodine bath for 10-15 min. Formaldehyde, up to 2,000 ppm for 15 min, can be used to treat eggs for the control of fungi.

e. Epizootic Ulcerative Syndrome (EUS)

EUS has a long history starting as early as 1975 in Australia and progressed in South East Asian countries reaching in 1992 in Srilanka and 1994 in Pakistan. Infection cause damage in a wide variety of fishes but Chinese carp and tilapia is comparatively resistant.

Channa stariatus (striped snakehead) an economically important species is highly affected. Infection starts as a red spot in the skin then it eventually becomes an ulcer. As it progresses, the ulcerative area gets eroded. Infected *Channa striatus* swims with a disintegrated caudal peduncle and eroded head due to the high general resistance to infections. The smaller freshwater fishes such as minnows die much before the infection can erode any organs. It also affects other *Channa* sp. and Indian major carps. Ulcers can be in any place of the body. The infection is common in freshwater and brackish water but not in marine fishes. It infects cultured species, wild species and large water bodies where there is no proper management.

Viruses were isolated from the EUS infected fish in South East Asian countries like Malaysia and Indonesia. Isolated viruses belong to rhabdovirus, reovirus and birnavirus.

But most often only fungus was isolated indicating the consistency of the fungal aetiology in EUS. The fungus is identified as *Aphanomyces invadans.* Regional council of FAO opined that EUS should have fungal hyphae in histological sections of underlying muscle of the ulcer. Grocott's stain is used for fungal staining. It gives black colouration in the fungal hyphae. *A. invadans* has the ability of penetrating the body of the fish using the proteolytic enzyme

It causes mortality by two means

1. By invading fungus
2. The invaded area is exposed to the environment resulting in osmoregulatory failure.

Control Measures

In the initial stage, application of low and sodium chloride treatment was found to be effective. CIFA produced a mixture of CIFAX, for controlling EUS but not much standardized.

General control measures include treating the water with chlorine before letting in to the pond. Avoid pumping water from large water body where there is no proper water quality management. Prevent the wild entry of fishes. Prevent the entry of birds as birds act as indirect vectors. Recovered infected fish is get immunity against this disease. In later stage, there are no effective control measures.

B. Fungal Disease of Shellfish

b.1. Larval Mycosis in Shrimps

Affected larvae appear opaque followed by sudden mortality. Larval stages are highly susceptible. The infection is caused by *Lagenidium* sp. *Sirolpidium* sp. and *Haliphthoros* sp. These are filamentous non-septate and coenocytic fungi. Once infected the larval mycelium ramifies into various parts of the body. *Fusarium* also causes infection in penaeid shrimps. Black gill disease of larvae is caused by *Fusarium.* Other fungi *viz. Saprolegnia* and *Leptolegnia* also cause dark necrotic lesions on shrimp exoskeleton and cause gradual mortality.

Chapter 6

Strategies for Health Management

While the diseases are difficult to be completely removed from aquaculture, it is possible to control them and prevent them from frequent incidences. As the disease in aquaculture is a result of interaction between host, pathogen and environment, controlling all the three factors become important. Keeping the water quality is the most important step in the drive against diseases. Controlling the entry of obligate pathogen to the culture systems and reducing build up of opportunistic pathogens are also equally important. Protecting the immune status of the host is also important which could increase their resistance of the fish or shrimp to infectious diseases. Apart from proper nutrition, several prophylatcic measures are required in achieving disease free aquaculture.

Selection of a species that can relatively withstand infections in comparison is one of the potential method of disease control in aquaculture. Selection of healthy larvae or juveniles and avoiding contamination during culture can also be used for effective disease control. Disease avoidance by using specific pathogen free (SPF) stock is a very useful method of disease prevention. Specifically, quarantining the stocks that are imported or transferred over a wide geographic area has to be practised.

Following methods also help in controlling the disease.

1. Use of closed or semi-closed recycle system.
2. Construct reservoirs for storing water without directly taking from the sea.
3. Minimise water exchange.
4. Treat reservoir water (chlorine at the rate of 30 ppm calcium hypochlorite - 60 per cent active ingredient) before use in the ponds,

5. Prevent entry of wild shrimps, crabs and fish in to the ponds.

6. In case of a disease outbreak, disinfect contaminated water before discharge.

7. Maintain good pond preparation by drying pond bottom and removing top layer of the sediment.

8. Avoid over stocking.

9. Maintain good water quality

10. Feed nutritionally balanced diet at the required quantity avoiding excess feed.

Infectious disease is considered to be the single most potentially devastating problem in shrimp culture. Selection of a species that can relatively withstand infections in comparison is one of the potential methods of disease control in penaeids. Selection of healthy larvae and avoiding contamination during culture can also be used for effective disease control. Disease avoidance by using specific pathogen free (SPF) stock is a very useful method of disease prevention. Specifically, quarantining the stocks that are imported or transferred over a wide geographic area has to be practised.

Conventional methods for controlling aquatic animal pathogens, such as chemotherapy, are less effective in managing newly emerging pathogens. Molecular biotechnological techniques has now acquired an increasingly important role in controlling the disease incidences by improving the effectiveness of screening and detection of pathogens, elucidation of pathogenicity, development of effective control and preventive measures, and treatment of diseases. The improvement of diagnostic methodologies for prevention, management and control of disease in cultured shrimp has emerged as one of the most important applications of biotechnology. Unlike traditional chemotherapeutic methods, which have been handicapped by development of pathogen resistance, the new technologies provide an opportunity for prophylactic intervention to minimize disease outbreaks.

Following methods are suggested by various workers towards controlling the diseases in shrimp farming especially the viral diseases:

1. Early and effective detection of pathogens using improved diagnostic methodologies to screen and quarantine infected population to prevent the spread of the pathogens.

2. Use of biosecured, closed or semi - closed recycle system with reduced water exchange and increased water – reuse culture systems.

3. Maintain good water quality and treat water before use especially for hatcheries. Disinfection with 30 ppm calcium hypochlorite (containing 65 per cent active chlorine) for 12 hours followed by neutralisation with soda ($Na_2S_2O_3$ @ 30 g/m^3) and vigorous aeration could be used to treat the incoming water.

4. Construct reservoirs for storing water without directly taking from the sea and treat reservoir water (chlorine at the rate of 30 ppm calcium hypochlorite - 60 per cent active ingredient) before use in the ponds,

5. Bio-security approach for preventing pathogen introduction by screening entry of infected hosts *i.e.* wild shrimps and non-host biological carriers like crabs, fish or birds in to the ponds and by avoiding the use of inanimate objects contaminated with pathogens. Stray dogs should be kept away from the vicinity of the ponds, as they can also cause spread of virus.

6. In case of a disease outbreak, disinfect contaminated water before discharge. Regular treatment of hatchery effluents also should be practised to prevent contamination of surrounding waters.

7. Maintain good pond preparation by drying pond bottom and removing top layer of the sediment. Application of 100 ppm CaO (burnt lime), exposure of pond bottom to sunlight for complete drying and ploughing of the soil are good measures of pond preparation after each harvest.

8. Avoid over stocking, optimum stocking density will reduce the stress on the animals and reduce the chances of water quality deterioration

9. Production of specific pathogen free (SPF) shrimps and the development of specific pathogen resistant (SPR) strains are two potential approaches, which are of immense use in the broodstock management programmes. SPF animals are produced by selecting animals free of known and detectable pathogens and raising them under controlled and strict sanitary conditions. The SPR animals are developed through selective breeding of animals known to be less susceptible to specific pathogens. These concepts are now being used in countries like the USA, Venezuela and French Polynesia with *P. vannamei* and *P. stylirostris.* The main benefit of this method is production of high health (HH) post-larvae, free of, or resistant to known pathogens. Reports indicate that *Penaeus merguiensis* has a higher resistance to WSSV than *P. monodon.* The selection of animals with reinforced non-specific defence or enhanced tolerance of external infection challenges is also a promising approach for controlling the disease outbreaks. However, in certain cases, it is feared that many SPF stocks, which have not been exposed to specific or general pathogens, perform poorly when exposed to pathogens in the field.

10. Feed nutritionally balanced diet at the required quantity avoiding excess feed.

11. Use of immunostimulants and vaccines to harness the specific and non-specific defence mechanisms of shrimps. This technique has considerable potential for health management in shrimp aquaculture. Biotechnological innovations are being developed in the field of immunostimulants and modulators in an effort to reduce shrimp susceptibility to disease.

12. Development of co-operative disease control strategies. co-operative program involving farmers, health care specialists, scientists and government agencies to track the movement of a potential pathogen, for notification of neighbouring farms of an infection and proper disinfection of the infected stock and water before discharge

13. Locate hatchery away from the shrimp farm to prevent cross contamination of hatchery source-water and air borne contamination of pathogens from the infected farm.

14. Use only virus-free broodstock to prevent vertical transmission of the viral diseases in particular.

15. Avoid importing of broodstock and larvae. This could increase accidental introduction of potential pathogens across the borders.

16. Maintain broodstocks of different sources in separate holding tanks and rear the larval population in separate batches to prevent cross contamination.

17. Do not use trash fish and shellfish to feed broodstock which can act as carriers of viral pathogens. In case of feeding with trash fish, care has to be taken for properly cooking the feed before use.

a. Vaccination

Basis of Fish Vaccination (The Immune Response)

The immune system is to protect the fish from bacteria, virus, or any foreign antigen (protein). Therefore, before attempting any vaccination strategy, it is important to determine when the immune system is both morphologically and functionally mature. Fish immunology has a more recent history than human and veterinary immunology but the techniques used are similar. However, methods of administering vaccines to fish differ and are dependent upon species, pathogen, temperature and environment.

Innate Immune System

Innate mechanisms require no previous exposure to the particular agent. This includes: physical barriers such as skin and mucus layers, specialized cells such as macrophages and natural killer cells and particular soluble molecules such as complement and interferon.

The first lines of defense of fish, which have against foreign agents, are mucus and skin, which contain immune-reactive molecules (*i.e.*, lysozyme, complement and immunoglobulin). Apparently, antibody is not produced in the serum but rather produced locally by mucosa-associated lymphoid tissues, which are sub-divided into gut, skin and gill. Non-specific cells of the fish immune system include monocytes or tissue macrophages, granulocytes (neutrophils) and cytotoxic cells.

As far as the complement is concerned, duplication and diversification of several complement components is a striking feature of bony fish complement systems. Recent studies have also confirmed the presence of functional homologues of mammalian cytokines in fish.

The Adaptive Immune System

Fish are a heterogeneous group divided into three classes: Agnatha (jawless fish such as the hagfish and lampreys), Chrondrichthyes (cartilaginous fish such as sharks, rays and skates) and Osteichthyes (bony fish). Fish above the level of the Agnatha display typical vertebrate adaptive immune responses characterized by immunoglobulins, T-cell receptors, cytokines, and major histocompatibility complex molecules. However, the immune system of fish is quite different in its efficiency and complexity from that of higher vertebrates. Acquired immunity in fish includes both humoral and cell mediated response. The cell-mediated response in fish is similar to that in mammals and relies on the presence of antigen results in a cascade of events that includes cytokine production that regulates or enhances the cellular response.

Most generative and secondary lymphoid organs in mammals are also found in fish, except for lymphatic nodules and bone marrow. The anterior portion of teleost fish (modern branch of bony fishes) kidney is most likely the source of histocompatibility complex molecules that will later give rise to the B and T-cell development takes place in the thymus of all vertebrates based upon an assortment of criteria. In teleost fish, progenitor T-cells migrates from the kidney to the thymus for T-cell education (distinguishing self from non-self) and maturation (functional). B-lymphocytes originate and mature within the kidney; therefore the anterior region of the fish kidney is considered to be evolutionary of the marrow. B-cells of fish produce antibody when stimulated.

Types of Fish Vaccine Formulation

Bacterins

Most bacterial vaccines in aquaculture to date have been inactivated vaccines obtained from a broth culture of a specific strain(s) subjected to subsequent formalin inactivation. Bacterins stimulate the antibody related portion of the immune responses (*i.e.*, the humoral immune responses). Whereas with some vaccine acceptable levels of protection are achieved with aqueous formulations administered by injection or immersion, for other bacterins, such as those devised for Salmonids against *Aeromonas salmonicida* sub sp. *salmonicida,* an acceptable level of protection can only be achieved by immunization with oil-adjuvanted bacterins delivered by injection.

Live Attenuated Vaccines

Live attenuated vaccines are composed of live microorganisms (bacteria, viruses) that have been grown in culture and no longer have the properties that cause significant disease. These vaccines potentially have many advantages in aquaculture. If the vaccinated fish shed the vaccine strain an effective dissemination of the antigen in the population would take place over an extended period. They also have the advantage that they stimulate the cellular branch of the immune system. Some live vaccines have been tested experimentally: *Aeromonas salmonicida, Edwardsiella tarda, E. ictaluri, P. damselae* sub sp. *piscicida.* However, problems concerning safety, persistence in the fish and in the environment, reversion to virulence, risk of spreading to non-target animals including wild fish, among others, must be resolved before the use of these live attenuated strains can be allowed in the field. At present, only an *E. ictaluri* attenuated live vaccine has been licensed in the USA to be used by bath in -9 day old fish to prevent ESC of catfish.

DNA Vaccines

DNA vaccines are composed of a particular portion of genetic material that can, after being incorporated into the animal, produce a particular immune-stimulating portion of a pathogen (*i.e.*, antigen) continuously, thus providing an "internal" source of vaccine material.

DNA vaccines have theoretical advantages over conventional vaccines: in mammals, the specific immune response after DNA vaccination encompasses antibodies; T-helper cells and cytotoxic cells. However, before DNA vaccines are applied in commercial enterprises in aquaculture, safety for the fish, environment and consumer have to be addressed. As the DNA-sequence encodes only a single microbial gene, there should be no possibility of reversion to virulence, which is a critical factor in relation to environmental safety in aquaculture.

It has been demonstrated that DNA vaccination induces a strong and protective immunity to some viral infections in fish, particularly the Rhabdoviruses infecting rainbow trout and Atlantic salmon, and also for channel catfish herpes virus infection.

Polyvalent Vaccines

The ideal vaccine formulation is a polyvalent vaccine, which protects simultaneously against the majority of the diseases to which a particular fish species is susceptible.

In addition, these polyvalent vaccines must cover all the main serotypes of each pathogen existing in a particular geographical area. Examples of the efficacy of polyvalent vaccines are those used in salmonids and turbot in which polyvalent vaccines give similar or superior protection than the respective monovalent vaccines. However, care must be taken in the formulation of polyvalent vaccines because the problem of antigen competition can occur, especially when these vaccines are administered by injection.

Route and Strategy of Administration

Fish are cold- blooded animals with a body temperature that equals their surrounding. Depending upon fish species and temperature, vaccination must be performed within a certain minimum period before the risk of their exposure to pathogens. In addition to temperature, stress caused by environments, crowding, handling and transport, can induce immune suppression and be a limiting factors for vaccine efficacy. Fish are commonly immunized by three procedures: intraperitoneal injection (ip), immersion in a diluted vaccine solution (short or long bath), or oral administration of the vaccine.

Although these methods have different advantages and disadvantages with respect to the level of protection, side effects, practicality and cost-efficiency, it is widely accepted that only the injection and immersion routes give enough protection to be used as the primary route of fish immunization in commercial production.

For oral vaccination, research has been focused on protecting the antigens from digestion and decomposition during passage through the stomach and anterior part of the gut. However, promising results have been obtained using encapsulation of antigens in alginate or polylactic glycolic acid microparticles. From the economic stand point, oral vaccination is the ideal route to be employed in a vaccination program which requires one or more booster immunizations.

Current Status of Fish Vaccines

Bacterial Vaccines

Vaccination plays an important role in large-scale commercial fish farming and has been a key reason for the success of salmon cultivation. In addition to salmon and trout, commercial vaccines are available for channel catfish, European seabass and seabream, Japanese amberjack and yellow tail, tilapia and Atlantic cod. In general, empirically developed vaccines based on inactivated bacterial pathogens have proven to be very efficacious in fish.

Furunculosis (Ulcerative disease of goldfish, *Aeromonas salmonicida*)

Furunculosis is diseases of fish caused by *Aeromonas salmonicida* subsp. *salmonicida* and, it can also affect fish from fry right through to brood stock, and the disease is often triggered by sharp rises in water temperatures combined with changes in fish physiology such as modification or spawning.

Although many furunculosis bacterins have been developed and commercialized since 1980, to be used in Salmonids by injection, immersion or the oral route their efficacy has been questioned because of the lack of repetitive results and/or the short protection period. The best results in terms of protection have been reported in Salmonids with the mineral oil-adjuvanted vaccines but it adherent to the viscera and a reduction in weight gain. To avoid these drawbacks, new nonmineral oil-adjuvanted vaccines have been recently developed and are now on the market.

Polyvalent vaccines, for Salmonids incorporating different *Vibrio* species and *A. salmonicida* as an antigens, are also available. DNA vaccines also were employed experimentally as safe live vaccines with a high level of success against furunculosis but their approval for use in the field has not yet been forthcoming.

Vibriosis

Vibriosis is one of the most important groups of bacterial diseases of marine fish with a world wide distribution. Within the genus Vibrio, the species causing the most economically serious diseases in marine culture are *Vibrio anguillarum*, *V. ordalii, V. salmonicida* and *V. vulnificus* biotype 2. *Vibrio anguillarum*, which is the cause of Vibriosis, has up to 23 O serotypes (O1-O2) are known only serotypes O1, O2 and to a lesser extent, serotype O3, have been associated with mortalities.

Although there are a great number of commercial *Vibrio anguillarum* vaccines have been developed for use mainly by bath or injection, the majority of them includes in their formulations only O1, or mixture of serotypes O1 and O2a. However, different polyvalent oil-adjuvanted vaccines, including different combinations of *Vibrio anguillarum* with other pathogens, such as *V. ordalii, V. salmonicida, Aeromonas salmonicida, Moritella viscose* and infectious pancreatic necrosis virus, are also available on the market to be used for Salmonids by the intra-peritoneal route.

Enteric Septicemia of Catfish (*Edwardsiella ictaluri*)

Edwardsiella ictaluri is the entero-bacterium responsible for enteric septicemia of catfish, with channel catfish being the most susceptible fish species among the ictalurids.

The bacterium is gram-negative, motile, pleomorphic curved rod, causing a major problem during the summer months when water temperature are above 18-28°C.

The first commercial Bacterins for *Edwardsiella ictaluri* were licensed to be used by immersion or oral routes. However, *Edwardsiella ictaluri* is an intracellular pathogen for channel catfish; it is not unusual that killed vaccines have not been very successful. Recently, an attenuated O-antigen deficient *Edwardsiella ictaluri* strain has been developed which was safe and provided high long-lasting acquired immunity (for at least 4 months) following a single bath immersion in 9-14 days old channel catfish without booster vaccination. This modified live *Edwardsiella ictaluri* vaccine has been produced since 2000, by Intervet Inc., under the trade name AQUAVAC-ESCO, and constitutes the first licensed bacterial live vaccine in aquaculture formulated with an attenuated pathogenic strain.

Columnaris Disease (Saddle Back Disease, *Flexibacter columnaris*)

Columnaris disease is a sub-acute to chronic disease in natural infections of most fresh water fishes affecting mainly ictalurids, eels, salmonids, cyprinids, centrarchids and ornamental fish such as golden shiner and goldfishes.

Several vaccination experiments against *F. columnare* have been performed on several fish species using different routes of administration (*i.e.*, injection, bath and oral) but the results in field trials were inconsistent, possibly due to the intimate association of stress with the disease process. Therefore no commercial vaccines are available.

Enteric Red Mouth Disease (Yersiniosis)

Enteric red mouth disease is caused by *Yersinia ruckeri*, that is, facultative anaerobic, non-motile, non-spore forming and Gram-negative rod, is mainly a fresh water disease of Rainbow trout, although it can affect other fish species such as Atlantic salmon in the fresh water phase and occasionally even at sea. The common vaccine commercially available recently is formalin inactivated whole cell cultures of *Y. ruckeri* serovar I, Biotype 1 (Hagerman strain). Bacterial kidney disease (*Renibacterium salmoninarum*).

Bacterial Kidney Disease (BKD)

It is caused by the Gram-positive diplobacillus group *Renibacterium salmoninarum* which is a fastidious, aerobic, non-motile, non-spore forming, gram-positive short rod bacterium. Although vaccination trials using classical bacterins, recombinant vaccines or attenuated live vaccines have been reported and there is evidence that under some conditions Renibacterium elicits an immune response in fish, the protective ability of a vaccine in field conditions is questionable because of the intracellular nature and vertical transmission of the pathogen, as well as the possible immunosuppressive role of the protein p57. Recently, a commercial aqueous live vaccine developed by Novartis has been licensed under the name of "Renogen" for BKD prevention.

Mycobacteriosis (Fish Tuberculosis)

Mycobacteriosis in fish (or fish tuberculosis) is a sub-acute to chronic wasting disease known to affect nearly 200 freshwater and saltwater species. Although *Mycobacterium marinum*, which is slow growing, non-motile, gram positive and acid fast rods, is considered the primary causative agent of fish Mycobacteriosis., At present no vaccines are available to prevent this disease in fish.

Cold Water Disease or Rainbow Trout Fry Syndrome (RTFS)

Flavobacterium psychrophilum (*Cytophaga psychrophila* and *Flexibacter psychrophilus*) has been known as the causative agent of bacterial cold-water disease (BCWD) or peduncle disease in Salmonids since 1948. The same bacterium has been shown to be the agent involved in the rainbow trout fry syndrome (RTFS) since the decade of the 1980s.

Recent vaccination experiments performed with young rainbow trout demonstrated that only significant protection was achieved using oil-adjuvanted ip vaccines; however, this route is impracticable for the early life fish stages in which *F. psychrophilum* infections usually occur. In addition, no cross protection among serotypes was obtained. Although no commercial vaccines against this disease are available, some countries are using autogenous bacterins made from single farm isolates.

Pseudomonadiasis

Among the *Pseudomonas* species recovered from diseased fish (*P. chlororaphis, P. anguilliseptica, P. fluorescens, P. putida, P. plecoglossicida*), *Pseudomonas anguilliseptica* is considered the most significant pathogen for cultured fish. Recent research efforts led to the development of aqueous and non-mineral oil adjuvanted bacterins (including both major serotypes detected), which proved to be effective in experimental trials in gilthead sea bream and turbot.

Viral Vaccines

In spite of the amount of research performed, both in commercial companies and in academic organizations, few viral vaccines are licensed. As of today, all fish virus vaccines for sale are based upon inactivated virus or recombinant proteins. No live attenuated or DNA vaccines are currently licensed, but one DNA vaccine against IHN (Infectious hematopoietic necrosis) disease is being tested in controlled field trials in Canada. Today, most available virus vaccines for aquaculture are based on inactivated virus or recombinant subunit proteins. In activated/killed viral vaccines are generally not efficacious unless delivered by injection, and as relatively, high doses are needed to achieve protection, cost-effective inactivated viral vaccines are difficult to develop. Live viral vaccines have been tested with good results in fish and should be the optimal regarding protection, administration and price.

Infectious Pancreatic Necrosis

Infectious pancreatic necrosis is a viral disease caused by an aquatic Birnavirus. This virus is related to infectious bursal disease (IBD) of poultry and in some studies the two viruses were morphologically indistinguishable.

The virus can cause problems in both fresh water and in the seawater phase of fish rearing. It tends to be a disease of younger fish, but the carrier status can exist which can give challenges in the control of the disease, especially in deciding where to transfer fish. There is a vaccine available for Atlantic salmon in the UK under a Provisional Marketing Authorization (PMA).

Pancreas Disease (Salmon Pancreas Disease Virus)

Pancreas disease is caused by an alpha virus, Salmon pancreas disease virus, which is very closely related to the virus causing sleeping disease of Rainbow trout. Although the disease is being controlled by bio-security, it is still a risk for trout growers. There is a Salmon pancreas disease vaccine available under a PMA. But unlike all the other combination Salmon vaccines designed for administration in a single injection this has to be given separately from any other injectable vaccine. To date there is not yet any vaccine available for trout.

Fish Vaccines against Parasites

There is wide range of parasites in both wild and cultured fish stocks. Although parasitic diseases such as amoebic gill disease, white spot disease, whirling disease, proliferative kidney disease (PKD) and Salmon lice infestation create several problems in fish farming, no parasite vaccines are commercially available. In general, fish possess both humoral and cell-mediated defense mechanisms against many parasites and there are many reports on immunity/increased resistance among fish surviving natural parasitic infection.

Cultivation of parasites for potential killed or live vaccine is even more expensive than virus cultivation, as a host population rather than cell cultures are usually required. In addition to the high costs, the use of natural hosts for cultivation of parasite would create major problems with respect to safety documentation.

Therefore, identification and production of protective antigens is probably the most feasible strategy towards commercial parasite vaccines, at least for low cost vaccines.

Limitations in Fish Vaccine Development

The major goal of vaccination is to induce a specific long-term protection against a certain disease. It has been debated whether the effective long term protection of oil-adjuvant injection vaccines, is due to immunological memory in the fish or constant stimulation from the antigen depot. As the existing empirically developed vaccines can induce protection after a single administration and until the fish are harvested, less effort has been put into the investigation of the actual mechanisms behind the protection.

As with all veterinary vaccines, cost effectiveness in the field is an essential limitation to commercial fish vaccine development. The ideal viral vaccine for aquaculture must be effective in preventing death, be inexpensive to produce and license, provide immunity of long duration, and be easily administered. But fish generally need a large antigen dose compared with terrestrial animals and cost-effective inactivated viral vaccines have proven difficult to develop. In some species, even all types of injection vaccines (or even immersion vaccines) are simply too expensive.

In the past ten years, commercial vaccine products for fish have more often consisted of mixtures of multiple products, including two, three, four and five vaccine. Considering the fact that not all antigens stimulate a protective immune response, that antigens vary in their immune-dominance relative to each other and that the immune system of fish has a defined and limited capacity to respond to individual antigenic substances, it becomes increasingly difficult to formulate these complex mixtures into safe and effective commercial products.

The other limitation is many fish species are too vulnerable to handle the stress induced during the vaccination or may develop severe side effects post vaccination for this matter, Oral vaccination should be considered as the most desirable method for immunizing fish because it is non-stressful, user-friendly and is capable of easy administration to large numbers of fish. Most of the research on fish vaccines has been performed by pharmaceutical companies, and not much information is available as scientific publications.

Yet, in other species, the major disease problems may appear in the larval or fry stages, before the animal is large enough to be vaccinated or have even developed functional immune system. The apparent lack of maternal immunity in fish also limits the possibilities to protect offspring by parental vaccination.

Future Prospects of Fish Vaccination

During the past 20 years fish vaccines have become an established, proven, and cost-effective method of controlling certain infectious diseases in aquaculture worldwide. Fish vaccines can significantly reduce specific disease-related losses resulting in a reduction of antibiotics use. To achieve progress in fish vaccinology, an increase in the co-operation between basic and applied science (*i.e.*, between the immunologist/microbiologist and the vaccinologist) is needed. There have been greatly advanced in the completion of genomic sequencing of pathogens, the application of comparative genomic and transcriptome analysis. This would facilitate to open opportunities up to investigate a new generation of vaccines; recombinant subunit vaccine, virus-like particle, DNA vaccine, and vector-vehicle vaccine. Currently, such types of vaccines are being actively explored against various fish diseases which depend on biotechnology, affording numerous advantages over conventional vaccines, including ease of production, immunogenicity, safety, and multivalency in a single shot.

Improvement in oral immunization with biodegradable micro particle based vaccines to be used for booster vaccination, development of new non-mineral oil adjuvants lacking side effects, development of polyvalent vaccines and standardization of a vaccination calendar appropriate for each economically important fish species with molecular biology and modern technologies are combining to make possible novel approaches to vaccine development.

Since resolution of virus persistence is thought to be correlated with cell-mediated immunity, vaccines designed to augment the cellmediated immunity must be developed for fish. Approaches that are being considered include the use of cytokines in combination with subunit vaccines and the use of specific MHC-I inducer adjuvants with the vaccine.

There are a number of potential vaccines for many fish diseases in aquaculture and so many studies have been performed or are in progress to formulate vaccines to prevent these diseases.

b. Antibiotics

The World Health Organization (WHO) warns the misuse of antimicrobial medicines and new resistance mechanisms are "making the latest generation of antibiotics virtually ineffective", while at the 2013 G8 Summit, scientific ministers issued a statement calling antimicrobial resistance (AMR) "a major health security challenge of the twenty first century."

Antibiotic use is an integral part of intensive animal agriculture and aquaculture. Increased public concern about antibiotic resistance and the need to preserve the ever-diminishing arsenal of antimicrobials that work in humans for as long as possible, has brought about increased scrutiny of the use of antibiotics in animal agriculture – especially for prophylactic and growth enhancing purposes.

The mechanisms by which antimicrobial resistant bacteria, initially derived from food-producing animals, contribute to the emergent and increasing threat of antibiotic resistance in people are complex and varied. The main routes, bacteria can take to move from animals to humans include *via* food or other animal product contamination, occupational exposure for farm workers and fish keepers, abattoir workers, veterinary surgeons and health workers. Bacteria can also transmit through environmental contamination like manure containing resistant bacteria, resistance genes, and antibiotic residues, along with recreational pursuits like swimming and fishing. The prevention of build up of resistant bacteria in waterways as a result of fish farming practices, terrestrial agriculture run-off or sewage outflow surrounding fish farms is a major concern for the aquaculture industry.

Despite difficulties of measuring the exact contribution of animal agriculture and aquaculture to the overall development of antimicrobial resistance, a consensus is emerging around the need for everyone to use these powerful drugs carefully and responsibly, especially those deemed critically important for human health by the WHO.

With the explosive growth in production and demand for farmed seafood, how can the aquaculture industry lead the charge for responsible use of antibiotics without compromising food safety, the environment and human health, as well as animal health, welfare and productivity?

Current Use in Aquaculture

The type and amount of use of antibiotics in aquaculture depends on farming practices, different local and national regulations and government enforcement ability. Some countries have recently introduced stricter regulations in response to the global threat of antimicrobial resistance and consumer concerns about residues in their food. The majority of aquaculture production, however, takes place in countries with "permissive regulations" and limited environmental monitoring. The overall use therefore varies widely between countries according to estimates by Defoirdt *et al.* (2011) who found antibiotic use ranging from 1 g per metric ton of production in Norway to 700 g per metric ton in Vietnam.

Much of the debate about antibiotic resistance focuses on two specific issues; firstly the practice of delivering antibiotics in small sub therapeutic doses in feed as a means to enhance the growth or prevent disease in fish and animals. In Europe, this kind of non-therapeutic use of antibiotics was banned in 2006 in an effort to curb over-use and reserve antibiotics for medicinal use only.

While prophylactic use of antibiotics is rare in aquaculture, metaphylactic use – the practice of treating an entire population of fish orally in to even if only a small percentage of the animals are affected in common place. In these cases, the sick fish will usually not eat the medicine so the effect is really to protect the healthy ones until the sick fish die and the infection subsides. As a result, the infection is rarely completely cleared. Furthermore, this form of oral treatment lends itself to sub-therapeutic doses which can enable selecting for resistance in bacteria.

Secondly, the specific need to curb the use of antibiotics categorised by the WHO as critically important for human health. While only a few antibiotics are approved for aquaculture and exact data on use is hard to come by (see below) at least two critically important antibiotics – tetracyclines and oxolinic acid, a third generation quinolone – are in regular use in Chile and Europe respectively to control some specific diseases and bacterial infections. The Aquaculture Stewardship Council (ASC) has made the use of critical antibiotics an issue of non-compliance for their certification, which could help push the development of alternative treatments and vaccines.

The Certification Approach

Another market-based approach to address these issues is provided by farm level certification schemes. While only five per cent of farmed seafood is currently certified, their market share is growing and consumer awareness about these schemes is on the increase. A positive development is that the three largest certification schemes all address antibiotic use within their frameworks for responsible aquaculture.

In a research report by the sustainable fisheries partnership comparing the certification standards for tilapia administered by the ASC, Global GAP and Global Aquaculture Alliance's Best Aquaculture Practices (GAA-BAP), they found that all of them emphasised practices and protocols to reduce the probability of risk events. In addition to requiring farms to implement comprehensive fish health management plans, the certification schemes also demand an active role of veterinarians through prescription of medicines and regular visits. Furthermore, a basic requirement across all schemes for tilapia include requirements against the use of growth promotion and prophylactic use of drugs and antibiotics.

The Path of Least Resistance

The certification schemes are doing a good job at raising awareness within the aquatic veterinary community about "the use of medically important antimicrobial drugs in food-animal production, and the public health risks associated with antibiotic resistance." At a joint aquaculture and agriculture industry round table discussion in Oxford, UK in May 2014, participants including vets, food scientists, farmers and representatives of the food and animal health industries, agreed that the development of a 'replace, reduce, refine' strategy could help drive the responsible use of antibiotics in food-producing animals.

Such an approach can guide action at company level as well as direct national policy by:

☆ Replacing antibiotics use where they are currently the first line of attack, primarily through comprehensive disease prevention strategies and practices. This is a similar approach to what is required by the major aquaculture certification standards. The development of bacterial vaccines can play a major role in replacing the need for antibiotics. This has been demonstrated through a steep decline in antibiotic use in salmon production in Norway, UK and Ireland after, for example, the introduction of the furunculosis vaccines.

☆ Reducing the use of antibiotics that are commonly deployed especially those critically important to human health by for example, better use of diagnostic and screening tools and standardised methods of measuring the use of antibiotics. Furthermore, use might be reduced through increased veterinary oversight and by addressing the dubious provenance and suspect purity of antibiotics available in certain markets.

☆ Refining the use of antibiotics to ensure intelligent and effective deployment (*e.g.* where they are needed to treat sick animals), including optimising recommendations on dosage, duration and routes of administration, as well as ensuring correct drug selection in light of concerns about antimicrobial resistance. There are only a few antibiotics licensed for use in aquaculture production. This presents an increased risk of inadvertent overuse lead to overuse of those few selecting for even more rapid resistance.

The certification standards have set aquaculture on a positive course, but unless the mainstream industry substantially changes its practice in the next few years, it is likely that governments in response to public concern will introduce regulative restrictions on antimicrobial use. It is also probable that if strict reductions in antimicrobial use were enforced without also introducing changes to standard farming practices, there could be severe consequences for the industry, fish and consumers, including increased disease and reduced productivity. The opportunity now is therefore for aquaculture practitioners, the industry and veterinarians to define and achieve more responsible antibiotic use at the farm level.

c. Immunostimulants

An immunostimulant is defined as a chemical, drug, stressor or action that enhances the innate or non-specific immune response by interacting directly with cells of the system activating them. Immunostimulants can be grouped under chemical agents, bacterial preparations, polysaccharides, animal or plant extracts, nutritional factors and cytokines. List of pathogen successfully controlled by using immunostimulants exposure in fish/shrimp like bacteria such as *Aeromonas hydrophila, A. salmonicida, Edwardsiella tarda, E. ictaluri, Vibrio anguillarum, V. vulnificus, V. salmonicida, Yersinia ruckeri, Streptococcus* spp.; virus such as infectious hematopoietic necrosis, yellow head virus, viral hemorrhagic septicemia and parasite *Ichthyopthirius multifiliis.*

Immunostimulants are dietary additives that enhance the innate (non-specific) defense mechanisms and increase resistance to specific pathogens. There is no memory component developed and duration of the immune response is very short. Immunostimulants are chemical substances which activate leukocytes. Adjuvant (FCA) is one of the first immunostimulants used in animals to elevate the specific immune response, and it has also been successfully used in conjunction with injection of fish bacterins. So far glucans, which are polymer of glucose found in the cell walls of plants, fungi and bacteria appear to be most promising of all examined in fish and shrimp and oral application found to be the route of choice. Use of these different types of immunostimulants is an effective means to increase the immune competency and disease resistance of fish and shellfish.

Concept of Immunostimulants

The infectious disease-outbreaks have emerged as constraints for the development of aquaculture. The occurrences have spread through the uncontrolled movement of live aquatic animals resulting in the transfer of pathogenic organisms among the countries. Antibiotics and chemotherapeutics have been used to prevent or control bacterial infections in aquaculture for about 20 years. Unfortunately, the use of antibiotics for treatment is not successful and sustainable due to increase in antibiotic-resistant bacteria, negative effect on the indigenous microflora of juveniles or adult fish, and the accumulation of antibiotic residues in fish tissue and environment causing human and animal health issues. Vaccination is an effective prophylactic treatment for infectious diseases in fish culture, but it may be very expensive and stressful to fish. A single vaccine is effective against only one specific type of pathogens, but limits the effectiveness for wide range of pathogens due to the complex antigenic structure. Therefore, the needs to look for alternative techniques with eco-friendly disease-prevention have been taken into account.

The alternative technique to prevent the diseases has been proposed that the strengthening of fish immune systems through the application of immunostimulants is one of the most promising methods. Immunomodulation by contrast, is a consequence of a change in the number or function of the cells involved in the immune response. The most proven effect of immunostimulants is to facilitate the function of phagocytic cells and increase their bactericidal and fungicidal activities. Immunostimulants can promote recovery from immunosuppressive states caused by any form of stress.

Fish will either survive if they successfully fight against the infection of pathogens, or die if they may not be successful in preventing the spread of the infection. The outcomes of survival or death are largely depended on the efficacy of the immune system to combat the initial infection or the spread of the pathogens. The immune system of fish can be grouped into acquired immunity (specific) and innate immunity (non-specific). Both use physical, cellular and humoral mechanisms to protect against infectious pathogens. The specific immune systems recognize specific antigen on a pathogen, and provide a protection against that specific pathogen. Non-specific immune systems provide a set of protective mechanisms that are inherently available for immediate protection against a wide variety of pathogens.

An immunostimulant is a chemical, drug, stressor, or action that elevates the non-specific defense mechanism or the specific immune response. Immunostimulants are practiced in aquaculture as a means to overcome the immunosuppressive effects due to stressors or might be used as a prophylactic treatment for expected seasonal outbreaks of known endemic diseases or a suppressive treatment for latent or sub-lethal pathogens. Immunostimulants can also promote recovery from immunosuppression states caused by stress. Various types of immunostimulants evaluated in fish and shrimp are summarized in Table 1.

Table 1: Immunostimulants Evaluated in Fish and Shrimp

Synthetic chemicals	Levamisole, FK-565 (Lactoyltetrapeptide from *Streptomyces olvaceogriseus*), Quaternary ammonium compounds (QAC).
Biological substances	
1) Bacterial derivatives	MDP (Muramyl dipeptide from Mycobacterium species), Lipopolysaccharide (LPS), Freund's complete adjuvant (FCA), EF203 (fermented egg product), Peptidoglucan (from *Brevibacterium lactofermentum* and *Vibrio* sp.), *Clostridium butyricum* cells, *Achromobacter stenohalis* cells, *Vibrio anguillarum* cells (Vibrio vaccine)
2) Yeast derivatives	ξ-1, 3 glucan, β-1, 6 glucan
3) Nutritional factors	Vitamin C and E, n-3 fatty acid
4) Hormones	Growth hormone, Prolactin, tri-iodothyronine
5) Cytokines	Interferon, Interleukin
6) Polysaccharides	Chitosan, Chitin, Lentinan, Schizophyllan, Oligosaccharide
7) Animal and plant extracts	Ete (tunicate), Hde (Abalone), Glycyrrhizin, Firefly squid, Quillaja, Saponin (Soap tree)
8) Others	Lactoferrin, Soyabean protein, Quil A, Spirulina, Achyranthesaspera (Herb), *Mucor circinelloides* (Fungi)

Vaccination is an important tool in preventing infectious disease in humans and animals and both passive and active vaccinations are extensively employed in fish. It is a term that should strictly be applied only when the purpose is long lasting protection through immunological memory. A vaccine targets the specific immune response. It requires primary challenge with antigen and is dependent upon the clonally derived lymphocytes subsets to be implemented. However, most commercial vaccine usually enhances resistance to only one or two specific pathogens and confers only a temporary resistance to disease. Immunostimulants by contrast, can boost immunity to a wide variety of pathogens, thus are nonspecific. Immunostimulations can be achieved in a more general sense by, for instance, targeting complement activation, phagocytosis and cytokines secretion, without necessary or purposefully requiring a specific response to a defined antigen. Examples include zymosan, glucans and lipopolysaccharides and these are best called as true immunostimulants. Comparisons of characteristics of vaccines vis a vis immunostimulants are given in Table 2.

Table 2: Vaccines vis-a-vis Immunostimulants

Sl.No.	Vaccine	Immunostimulants
1.	Prophylactic for long duration with only one or two treatments.	Prophylactic for short duration, require more treatments.
2.	Efficacy of vaccine is excellent	Efficacy of immunostimulants is good
3.	Limited spectrum of activity.	Wide spectrum of activity
4.	No toxic side effects	No toxic side effects
5.	No accumulation of toxic residues	No accumulation of toxic residues
6.	No environmental impact	No environmental impact
7.	Enhance specific and nonspecific immune response	Mainly enhance nonspecific immune system of larvae before specific immune system matures.
8.	Difficult to vaccinate larvae of fish and shrimp	Easy to vaccinate larvae of fish and shrimp
9.	Costly	Cost effective

Application of Immunostimulants Used in Fish and Shrimp Synthetic Chemicals

Levamisole is an antihelminthic chemical compounds used to treat the nematodes infection in human and animals as well. It can stimulate immune response *in vitro*. Levamisole enhanced phagocytic activity, the NBT reaction and increase antibody producing cells oral administration of levamisole increased the number of leucocytes, lysozyme activities in serum and the stimulated NBT reduction and phagocytic index of phagocytic cells. However, no differences were found in the levels of hematocrit, leucocrit or immunoglobulin using levamisole in rainbow trout. It has been observed that rainbow trout exposed to a bath treatment containing 5, 10, 25 µg/ml levamisole for 2 hrs period showed resistant to *Y. ruckeri*.

Biological Substances

Bacterial Derivatives: MDP (Muramyl dipeptide)

MDP (N-acetylmuramyl-L-alanyl-D-isoglutamine), obtained from Mycobacterium. Intra peritoneal injection of rainbow trout with MDP-Lys increased the phagocytic activities, respiratory burst and migration activities of kidney leucocytes as well as resistance of the fish to *A. salmonicida*.

Table 3: A Comparison of Characteristics of Vaccines and Immunostimulants

Name	Company	Type	Mode of Action	Instruction for Use	Specific Supporting Literature	Recommended Level of inclusion (kg/ton)
Immustim	Immundyne, USA	β (1,6) branched β (1,3) glucan from yeast	Activates macrophages	4 days-yes 3 days-no	General on similar products	0.5-12.5
Macroguard	Biotec-Mackzymal, Norway	β (1,6) branched β (1,3) glucan from yeast	Activates macrophages	-With vaccines in injection - in feed <6-8 weeks cont.	Yes, on salmonids and tiger shrimp	1.0
Vitastim	Taito and Company, Japan	β (1,6) branched β (1,3) glucan from fungi	Activates macrophages	As feed ingredient	Yes, on shrimp and fish	1.0 50 mg/kg body weight
Aqua-Mune	Park Tonks	β (1,6) branched β (1,3) glucan from baker's yeast	Activates macrophages	As feed ingredient	No	1.0
Penstim	AURUM Aquaculture Ltd, USA	Beta-glucan	Activates macrophages	Immersion for larvae and feed ingredient for larger animals	-	2.0 to 5.0 in feed
Laminarn	Pronova, Noway	β (1,6) branched β (1,3) glucan from brown algal laminariae	Activates macrophages	As feed ingredient	Yes	-
Calcium spirulan	Kelly Moorhead, Hawaii	Sulfated polysaccharide from spirulina	Inhibit viral envelop replication, inhibits virus penetration in host cell	Not known	Yes, on mammals	Not known

Contd...

Table 3–*Contd...*

Name	Company	Type	Mode of Action	Instruction for Use	Specific Supporting Literature	Recommended Level of inclusion (kg/ton)
SP 604	Alltech Inc, USA	Premix of mannan (Biomos), Cr and Se yeast, probiotics	Multiple: activates macrophages, source of trace minerals			
Agrimos	Agrimerica, USA Santel, France	Mannan based oligosaccharide from	Good substrate for lactic bacteria, occupies binding sites of pathogens in gut, stimulates immune response	As feed ingredient	Yes, on mammals	1.0
Elorisan	BUGICO, Switzerland	Organic silicon	Lower lipid permeability, protection of nervous tissues	Intraperitoneal injection	Yes	0.1 ml of 1 per cent solution
DS 1999	International Aquaculture Biotechnologies Ltd	Bacterin	Activates macrophage	Add directly in culture medium (larval culture)	Yes, field trails	0.5 in larval diets
Levamisole	Janssen Pharmaceutica, Belgium	Tetrahydro-6-pheny limidazolthiazole hydrochloride	Activates macrophages	As feed ingredient in bath	Yes, on mammal and fish	5-10mg/kg body weight
Lysozyme Hydrochloride	Belovo, Belgium	Lysozyme from hen's eggs	Kills/lyses bacterias	As feed ingredient	Yes, mammals (Cheese)	-
Lactoferrin	DMV international, the Netherlands	Lactoferrin from bovine milk	Binds Iron, makes it unavailable to pathogens	As feed ingredient	Yes, on mammals	-

Contd...

Table 3–Contd...

Name	Company	Type	Mode of Action	Instruction for Use	Specific Supporting Literature	Recommended Level of inclusion (kg/ton)
Selenium yeast	Alko, Finland	Selenium Yeast	Acts a yeast (activates macrophages) and source of selenium	As feed ingredient	-	1.0
Polypeptides fish hydrolysates	Tepual, Chile Sopropeche, France	Short peptides	Activates macrophages	As feed ingredient	-	-
Blood plasma	Harimex, The Netherlands	Blood serum (immunoglobulins, glycoproteins)	Activates macrophages, Binds on gut bacterial receptors	As feed ingredient	Yes, on mammals	50-100
Fish oils	Fish oil producer	Omega-3 fatty acids	Lower prostaglandin E-2 production (Immunodepressor), increase membrane fluidity	As feed ingredient	Yes, on mammals	10-100

LPS (Lipopolysaccharide)

LPS is a cell wall component of gram-negative bacteria. It was reported that LPS effective in preventing *A. hydrophilla* disease and stimulating innate immune response of rainbow trout. LPS can stimulate phagocytosis and the production of superoxide anions in Atlantic salmon macrophages and LPS can stimulate B-cell proliferation and enhance macrophage phagocytic activity in red sea bream *Pagrus major*. Also, LPS stimulates the production of macrophage activating factor in goldfish lymphocytes. These substances are very potent even in very low doses and may occur as contaminants in bacterin preparations and used in fish immunizing programmes. LPS stimulate hemocytes proliferation; enhance phagocytic activity as well as the microbicidal activity of shrimp.

FCA (Freund's Complete Adjuvant)

FCA is a mineral oil adjuvant containing killed *Mycobacterium butyricum*, increase the immune response in fish. FCA can increase respiratory burst, phagocytic and NK cell activity of leucocytes in rainbow trout protect against *V. anguillarm* infection. In contrast, yellowtail injected with FCA did not show increased resistance to *P. piscicida* infection, although the adjuvant effect of FCA on a *P. piscicida* vaccine was observed in fish.

Vibrio Bacterin

Vibrio anguillarum bacterin (inactivated whole cell vaccine) is the most successful vaccine for salmonid fish, administered through injection, oral and immersion methods. Immunostimulation of *V. anguillarum* bacterin was seen in fish and shellfish. In black tiger shrimp the migration of hemocytes treated with vibrio bacterin can be increased. Further, vaccination of rainbow trout with attenuated *V. anguillarum* stimulates protection against *A. salmonicida* challenge.

Clostridium butyricum Cells

C. butyrium bacterin can enhance the resistance to vibriosis in rainbow trout by oral administration by leucocyte activation, including phagocytosis and increased superoxide anion production. *C. butyricum* shows immunostimulatory effects like stimulation of macrophages and NK cells and improves further protection against Candida infection.

Achromobacter stenohalis Cells

A. stenohalis is a gram-negative aerobic organism which has been isolated from sea water. Inactivated *A. stenohalis* can enhance immune responses of kidney cells, complement activity and increase protection against *A. salmonicida* challenge. The LPS of this bacterium activates mouse macrophages and B-lymphocytes.

EF203

EF203 is a fermented product of chicken eggs and oral administration of it to rainbow trout stimulates the activity of leucocytes such as phagocytosis and chemiluminescence and increases protection against *Streptococcus* infection.

Yeast Derivatives

Glucans, long chain polysaccharide extracted from yeast, are good stimulators of non-specific defence mechanism in animals including fish and shellfish like phagocytic activity and protection against bacterial pathogens. Several types of glucan has been investigated in fish such as yeast glucan, peptide-glucan β -1,3, glucan (VST). Yeast glucan (β 1-3- and β1-6-linked glucan) and β-1,3glucan (VST) is derived from cell walls of baker's yeast like *Saccharomyces cerevisiae* and *Schizophyllum commune*, respectively. β-glucans comprised diverse group of polysaccharides of D-glucose monomers linked with β-glycosidic bonds. Cellular and non-cellular defense mechanisms are increased in activity after treatment with β-glucan like lysozyme activity, phagocyte activity, complement activity and bactericidal activity of macrophages. Dietary β-glucan administration increases resistance to infection. the highest antibody titre against *A. hydrophila* injected with β-glucan (100-1000 µg glucans/fish). In addition, intraperitoneal injection of β-glucan prepared from cell walls of *Saccharomyces cerelisiae* injected to Atlantic salmon showed increased resistance to *V. anguilllarm, V. salmonicidia* and *Y. ruckeri.*

Polysaccharides

Chitin and Chitosan

Chitin is a polysaccharide which constitutes the principal component of exoskeletons of crustacean and insect and cell walls of few fungi. It can stimulant macrophage activity and give resistance from certain bacteria. Chitosan, N-acetylated chitin can increased protection against *A. salmonicida* infection when injected with or immersed in chitosan solution in brook trout, *Salvelinus fontinalis.*

Lentinan, Schizophyllan and Oligosaccharide

Lentinan, Schizophyllan and Oligosaccharide can increase cellular and noncellular defense mechanisms like lysozyme activity, phagocyte activity and complement activity in fish.

Animal and Plant Extracts

Ete (Tunicate) and Hde (Abalone)

An extract from the marine tunicate, *Ecteinascida turbinate* (Ete) and a glucoprotein fraction of water extract (Hde) from abalone, *Haliotis discus hannai*. It can enhance the killing of tumor cells *in vitro* and inhibits tumor growth *in vivo*. Ete (Tunicate) can enhance the phagocytosis and increased survival of eel when injected against *A. hydrophila*. In addition, when rainbow trout injected with Hde against *V. anguillarum* infection showed increased survival along with enhanced phagocytic activites.

Firefly Squid

Firefly squid, *Watasenia scintillans*, can stimulate the immune system of rainbow trout such as the production of superoxide anion, potential killing activities by macrophages and the lymphoblastic transformation of lymphocytes *in vitro*.

Chinese medicinal herbs (*Astragalus membranaceus* and *Lonicera japonica*) extracts can be used as immunostimulants to enhance immune response and disease resistance of cultured fish species. The herbal immunostimulants such as *Emblica officinalis*, *Cynodon dactylon* and *Adathoda vasica* improved the immune system and reduced microbial infection in the goldfish *Carassius auratus* and similar work was also carried out on another ornamental fish *Poecilia sphenops* using herbal immunostimulants. Nile tilapia shows enhanced phagocytic activity after treated with Astragalus extract for one week.ginger extract to be very effective in enhanced phagocytic and extracellular burst activity of white blood cells in rainbow trout.

Nutritional Factors

Vitamin C

Vitamin C is involved in several physiological functions including growth, development, reproduction, wound healing, response to stressors and possibly lipid metabolism through its action on carnitine synthesis while administering in feed. Vitamin C (Ascorbic acid) is a co-factor in many biological processes including collagen synthesis and cellular functions related to neuromodulation, hormone and immune systems. It has been observed that higher levels of dietary vitamin C significantly increased the protection against *A. hydrophila*.

Vitamin E

Vitamin E can enhance specific and cell-mediated immunity against infection in Japanese Flounder. *Paralichthys olivaceus* and macrophage phagocytosis in fish such as channel catfish *Ictalurus punctatus* and turbot *Scopthalamus maximus*. Vitamin E deficiencies in trout result in reduced protection against *Y. ruckeri*.

Hormones

Growth Hormone (GH)

GH directly affects immune competent cells like macrophages, lymphocytes and NK cells. In fish, exogenous growth hormone (GH) has mitogenic activity on lymphocytes and activates NK cells and production of superoxide anions of leucocytes.

Prolactin

Prolactin also directly affects immune competent cells like macrophages, lymphocytes and NK cells. It can enhance the production of superoxide anions of leucocytes. prolactin helps in increased level of production of superoxide anion in rainbow trout by leucocytes.

Lactoferrin, consist of a single peptide chain with molecular wt. ~ 87,000 Da and posses 2 Fe-binding sites per molecule, most popular physiological fluids of mammals.

Cytokines

Cytokines are polypeptides or glyco-proteins which act as modulators in the immune System. Cytokines may be useful as powerful immunostimulants if their structures can be identified and recombinant molecules prepared.

Algal Derivatives

Laminaran is a β (1, 6)-branched β (1, 3)-D-glucan, a major component in sub-littoral brown algae, *e.g.* Phaeophyceae. Almost all B-(1, 3) D-glucan display poor water solubility which makes them less easy to handle than aqueous soluble laminaria. Laminaran obtained from *Laminaria hyperhorea* has immune modulatory effect on immune system as well.

Immunological System in Fish

The immune system is the system which continuously fights against the pathogen and give proper protection to our body. The two types of immune system are innate immune system or non specific immune system and acquired immune system or adaptive immune system. The essence of the immunological system of the vertebrates is to react and protect against the infections. Proper work of the immunological system of a fish involves different cells and organs. There are different factors that affect the immunological response of the fish. Inherent factors like health and age, extrinsic like temperature or changes in abiotic parameters them all together affect the health condition, and thus the response. Those changes cause in some cases stress, which if achieves high level generates an immunological system collapse.

Cells involving on the immune system are leucocytes or white cells. Those can be found in the blood stream or on tissues. Lymphoid tissues on fishes are thymus, spleen, anterior kidney and lymphoid tissues associated to mucus and intestine. The classification of leucocytes as in the vertebrates has been done following morphological criteria whereby various groups can be distinguished such as lymphocytes, granulocytes and macrophages. A short explanation of each group is done below in order to know the main characteristic and functions.

Lymphocytes

High differentiated cells with capability to respond on stimuli. The most common are mature lymphocytes with an irregular surface or border. Previous studies have defined lymphocytes as a high metabolically potential due to its high number of organelles in the cytoplasm as golgi apparatus, mitochondria, ribosome, and endoplasmic reticule. They are found in all over the body circulating on the blood stream and gathered on lymphoid organs and the quantity is very variable. The main function is to produce antibodies, immunological memory, and regulatory factors as lymphokines in response of the humoral and cell specific immune. Lymphocytes B are bone marrow derived while T is thymus derived. T cells are responsible for cell mediated immunity as well as providing assist to B lymphocytes; those last are responsible to produce antibodies against antigen.

Granulocytes

Occurrence and functions varies within species of fish, the origin is focus on the kidney tissues. In teleost, there are describe three based on morphology including neutrophil, eosinophil and basophil, the first being the most common. Granulocytes react responding in the presence of foreign material going into the body but without recognizing specific antigens. This kind of defending is called in non-specific defense mechanism, explained little forward. When the invasion is occurred those cells migrate and destroy the estrange particles by phagocytosis or just by killing by a cytotoxic response.

Macrophages

Based on literature many test has been done in several species of fish. After several tests it seems that macrophages can be use to evaluate the health of the fish as a kind of indicators. Those cells play an important role on killing pathogens as immune response. macrophages are the main phagocyte cells on fishes. The pathogens are killed by two ways, releasing toxic substances or by ingestion, known as phagocytosis. It involves producing ROS-reactive oxygen species or microbiocidal oxygen radicals. This generating activity is known as respiratory burst, and is not only produce by phagocytosis. Lymphokines can regulate macrophage functions like MAF, macrophage activating factor. The immunological system of fish can be divided in two branches depending on the functionality such as natural or nonspecific and acquired or specific. The non-specific immune system is considered to be the most important ones for immunonisation.

Immunostimulation of Non-specific Defense Mechanisms

Most immune-stimulatory compounds examined in fish and shellfish have been shown to have immune enhancing potential through heightening of nonspecific immune responses of the organisms. The nonspecific immune responses in contrast to the specific response does not require prior exposure to an antigen and consists of barriers such as skin, scales, lytic enzymes and phagocytic cells. The nonspecific immune response is also considered to be the first line of defence against invading pathogenic microorganisms and is the sole immunological mechanism by which invertebrates protect themselves from disease. In contrast to specific immunity, which only recognizes a particular antigen or pathogen, each component of the nonspecific response can recognize a broad array of foreign agents. It has been hypothesized that fish and shellfish are more reliant on nonspecific immune response. For these reasons, a large portion of the research on immunostimulation has focused on up-regulating the nonspecific immune response of the organisms.

It is well established that mononuclear phagocytes or macrophages plays a central role in the cellular part of the nonspecific defence mechanism of fish. Haemocytes and the prophenoloxidase (proPO) system are the primary defence mechanisms of shrimp. Both semi-granular and granular cells carry out the functions of the proPO system. Phenoloxidase is the terminal enzyme in the proPO activation system and is activated by lipopolysaccharides or peptidoglycans from bacteria and β-1, 3 glucan from fungi through the pattern recognition molecules.

Phenoloxidase activity has been detected in many species of penaeid shrimp such as Sao Paulo shrimp *Farfantepenaeus paulensis*, yellow leg shrimp *Fafantepenaeus californiensis*, tiger shrimp *Penaeus monodon*, blue shrimp *Litopenaeus stylirostris* and white shrimp *L. vannamei*. The activation state of these cells and enzyme systems are often used as measures of non-specific immunostimulation. Other measures employed for this purpose include cell migration, phagocytosis and bactericidal activity as well as changes in numbers of leucocytes and the activation potential of cells upon stimulation, as measured by oxidative radicals and enzymes. The nonspecific immune responses such as phagocytosis and the production of oxidative radicals are quickly activated by the immunostimulants and help to protect the host against a broad spectrum of pathogens.

Method of Administration

Immunostimulants potentiate the immunity of the host itself, enabling it to defend more strongly against pathogens. Several immunostimulants also stimulate the natural killer cells, complement, and lysozyme and antibody response of fish. There are mainly 3 ways to deliver immunostimulants including injection, immersion and oral uptake. Injection of immunostimulants can produce strong non-specific response but its costly affairs with lots of time and labour intensive as well, applicable only for large size of fish more than 10-15 g in body weight in intensive aquaculture system. It has been reported that injection has wide protection against a range of pathogens like intra-peritoneal injection with glucan injected to channel catfish shows increased in phagocytic activity reducing fish mortality challenge with *Edwardsiella ictaluri*. For small fish vaccination is impractical. Immersion produces less non-specific immune response, but more cost effective than injection, increase more stress to fish while handling, applicable in intensive culture system. Immersion method is very effective during acclimation of juveniles to ponds in field condition. Using immersion of levamisole showed increase in circulating leukocytes, phagocytic rate and increase protection against *P. damselae* sub sp. *piscicida* in European Seabass.

Oral ingestion produces good non specific immune response and can be the most cost effective method with economically viable. It is mostly suited for extensive aquaculture system. Immunostimulants powders are mixed with feed using a fish oil coating. Now a day, bioencapsulation method is also followed to immunize the fish larvae during their early larval stages with live fed organisms.

Timing of Administration

It is necessary to apply immunostimulants at the right time. The application of immunostimulants should be implemented before the outbreak of disease to reduce disease related losses. Effective dosage and exposure time will be further more complicated based on different culture systems with feeding regime. In Atlantic salmon injection with high dose of glucans @ 100 mg/kg led to absence of protection for 1 week, but maximum benefits only occurs after 3-4 weeks. Also, at low dose of injection @ 2-10 mg/kg, give protection only 1 week. Similarly, it has been noticed that increase in the number of NBT positive cells in African catfish fed with glucan or oligosaccharide over 30 days, but not over 45 days.

Mode of Action

The mode of action of immunostimulants is to activate the immune systems of organisms, to enhance the immunity level against invading pathogens. The approach is very diverse in nature or may be poorly understood and also depends on the type of immunostimulants, dose, route of administration, time and length of exposure.

Following are some of the mechanism of actions:

☆ Stimulators of T-lymphocytes- Levamisole, Freund's Complete Adjuvant (FCA), Glucans, Muramyl dipeptide, FK-565 (Lactoyl tetrapeptide from *Streptomyces olivaceogriseus*).

☆ Stimulates of B-cells- Bacterial endotoxions, Lipopolysaccharides.

☆ Macrophage activator- Glucans, Chitin and Chitosan

☆ Inflammatory agents including chemotoxins

☆ Cell membrane modifiers- Detergents and Sodium dodecyl sulphate, Quaternary ammonium compounds (QAC), Saponins

☆ Nutritional factors-Vitamin C and E, n-3 fatty acids

☆ Cytokines- Leukotriene, Interferon

☆ Heavy metals- Cadmium

☆ Animal and fish extracts- Mitogens

In general immunostimulants enhance the phagocytosis and bacterial killing ability of macrophage, complements, lymphocytes and nonspecific cytotoxic cells, resulting in resistance and protection to various diseases and invading microorganisms.

Detection of Immunostimulation

An increase in any characteristics such as phagocytosis, production of superoxide anions *etc.* in treated fish and shellfish over controls is evidence of immunostimulation. Following are some of the methods of detection of immunostimulation.

Haematocrit and Leucocyte Count

Leucocytes mediate nonspecific immunity. So raised leucocytes count with an essentially unchanged haematocrit is an indication of immunostimulation.

Phagocytic Activity

Phagocytosis is a common reaction of cellular defence and generally recognized as a central and important way to eliminate microorganisms or foreign particles. Phagocytosis can be assayed by incubating blood with a killed bacterial culture and examining stained smears for phagocytes containing bacteria.

The phagocytic activity are defined as phagocytic ratio (PR) and phagocytic index and are expressed as

$$PR = \frac{\text{Number of phagocytic cells with engulfed bacteria}}{\text{Number of phagoctes}} \times 100$$

$$PI = \frac{\text{Number of engulfed bacteria}}{\text{Phagocytic cells}}$$

Bactericidal Activity

Bactericidal activity can be assayed by incubating macrophages with a live bacterial culture and then washing off the supernatant liquid, lying the macrophages and examine the numbers of live bacteria.

Oxidative Radical Production

A major way in, which neutrophil granulocytes contribute to nonspecific immunity, is by the production of oxidative radicals. Nitro-blue tetrazolium (NBT) reacts with oxidative radicals producing a dark blue color and is used to identify neutrophils actively producing them.

Myelo-peroxidase Production

Activated neutrophils also produce myelo-peroxidase. The level of activation can be determined by incubating blood smears in an indicator reagent and examining cells under the microscope for degree of staining.

Immunoglobulin Concentration

Some serum immunoglobulins are humoral antibodies and therefore heighten specific immunity, many others regulate nonspecific immunity.

In vitro Measurement

An *in vitro* method for screening substances for immunostimulation. In essence finally divided pieces of rainbow trout spleen are maintained in a tissue culture medium with a test substance and after 4 days cell suspension are prepared. For neutrophils the cell suspensions are treated with NBT and examined by spectrophotometry; for phagocytes aliquots of the cell suspension are shaken for 15 min. with a suspension of fixed sheep erythrocytes and then smears are made for microscopy.

In vivo Measurement

In fish, specific immunity develops slowly and thus it is possible to assess immunostimulation by a challenge test with virulent bacteria, which rapidly kills large number of fish at a time. Any delay or reduction in mortality in treated fish compared to untreated group may be attributed nonspecific immunity systems.

Attributes of Immunostimulants

- ☆ Safe for the environment and human health, biocompatible and biodegradable
- ☆ Promote a healthy body status by triggering the immune system of the host
- ☆ Nontoxic to finfish and shellfish with no known side effects
- ☆ Enhance disease resistance against broad spectrum of pathogens
- ☆ Reduce mortality due to opportunistic pathogens
- ☆ Prevent viral diseases
- ☆ Enhance the efficacy of antimicrobial substances
- ☆ Enhance the efficacy of vaccines and antibiotics
- ☆ Cheap, ecofriendly and easily available.

Efficacy and Limitation of Immunostimulants

The use of immunostimulants can protect fish from several infectious diseases and decrease mortality rates by increasing fish resistance against infectious bacteria such as *Vibrio anguillarum, V. salmonicida, Aeromonas salmonicida* and *Streptococcus* sp., viral infections such as IHN (Infectious Hematopoetic Necrosis) and yellowhead (YHV) disease and parasitic infections such as white spot disease and sea lice; and immunostimulants do not increase resistance against *Renibacterium salmoninarum, Pseudomonas piscicida* or *Edwardsiella ictaluri* infections due to their resistances to phagocytosis and abilities to survive within macrophages. Use of immunostimulants in cultured fish result in macrophage activation, increased phagocytosis by neutrophils and monocytes, increased lymphocyte numbers, increased serum immunoglobulins, and increased lysozyme. Different immunostimulants have effectiveness for different life stages based on solubility and give a different degree of protection against the pathogens. An algal extract, laminaran, more soluble than the fungal and yeast glucans, has also proven to activate macrophages and to increase respiratory burst activity in anterior kidney leucocytes of Atlantic salmon. The laminaran has been promoted for its superior solubility compared with other β-1, 3-glucans, considering as candidate substances for diet application. Absorption of laminaran from water by yolk-sac larvae of Atlantic halibut, *Hippoglossus hippoglosus*, through the skin and intestinal epithelium suggests that it may have the potential to enhance immunity in early life stages before the development of acquired immunity.

However, effect of immuno stimulants on nonspecific immune mechanism is normally of short duration. It has been shown in salmon gave maximum leucocyte responses just 2 days after injection with M-glucan, and that yeast β -glucan gave an increase in respiratory burst activity 4-7 days after treatment. The protection in trout using glucans and chitosan was greatly reduced after 14 days of immune stimulation. The use of several immunostimulants for prolonged periods does not appear to provide additional advantages with respect to a single dose. After 29 days, the protection induced in blue gourami by injection of 20 mg/kg laminaran for 22 days, was not significantly different from the effect of a single injection of 20 mg/kg laminaran. But, most of the immune parameters such as leucocyte count, phagocytic ratio, phagocytic index, lysozyme activity, complement activity, serum bactericidal activity were significantly enhanced on 42 days after three i.p. injection of 10 mg of β-glucan/kg body weight of *Labeo rohita* fingerlings, which would lead to long-term protection in fishes. The effects of immunostimulants are not directly dose dependent, and high dose or overdosage may not enhance and may inhibit the immune responses, reported that the increase in respiratory burst activity of glucans-treated macrophages was maximal at glucan concentrations of 0.1-1 µg/ml, whereas at10 µg/ml no effect was seen and at 50 µg/ml glucan was inhibitory. Vitamin E is immunostimulatory in concentration of 50-300 IE/kg feed, but very high vitamin E levels (>1000-5000 IE/kg diet) fed for a prolonged time have an immunosuppressive effect. Genetics of the species, life history stage, and the culture environment all interact with the type and dosage of immunostimulants to contribute to the efficacy of the immunostimulatory substance.

Immunostimulants in Aquaculture Health Management

Immunostimulats have been extensively studied in fish and shellfish both at whole animal and on a cellular level. It has been used as prophylactics to control infections disease of animals and also playing the role of alarm molecules that activate the immune system. Fish and shrimp depends more heavily on nonspecific defence mechanisms than mammals and therefore immunostimulats play a vital role in health management strategies of aquatic organisms. There are at least 20 different compounds, including levamisole, lipopolysaccharides, glucan, vitamin C and E *etc.* that are used as immunostimulats, adjuvants and vaccine carries in fish. Among these compounds, glucan is one of the most promising stimulants for nonspecific defence mechanism and also the most studied immunostimulant in aquatic species. Glucan has been reported to enhance resistance against bacterial pathogens such as *Vibrio anguillarum, Aeromonas salmonicida, A. hydrophila* and *Yersinia ruckeri* in several species in fish such as the carp *Cyprinus carpio*, Atlantic salmon *Salmo salar,* rainbow trout *Onchorhynchus mykiss*, yellow tail *Seriola quinquradiata* and African catfish *Clarias gariepinus*.

Applications of β-1,3 glucan enhances the nonspecific cellular defence mechanisms of animals by increasing the number of phagocytes and the bacterial killing activity of macrophages in rainbow trout, Atlantic salmon, catfish, and carp and through production of superoxide anions by macrophages. In an Indian major carps, Labeo rohita, yeast glucan have been observed to enhance the phagocytic activity of leucocytes and stimulate generation of reactive oxygen species (ROS) in phagocytes.

In recent years, nucleotides and their metabolites have received heightened attention as potential immunomudulators. They play key roles in numerous essential physiologic functions, including encoding genetic information, mediating energy metabolism and signal transduction. The beneficial influences of oral administration of nucleotides on immune functions, vaccine efficiency or disease resistance has also been demonstrated in fish such as Atlantic salmon, coho salmon, rainbow trout, common carp, hybrid tilapia and hybrid striped bass.Dietary nucleotides are capable of enhancing the potential of the immune system in general to mount greater and more rapid specific responses, as compared to the primarily nonspecific capacity of phagocytes induced by glucan. dietary yeast ribonucleic acid at 0.4 per cent enhances phagocyte respiratory burst and protection of Labeo rohita juveniles by Aeromonas hydrophila. In addition, dietary nucleotides have also been observed to modulate gene expression, a phenomenon also confirmed in fish. However, the way in which dietary nucleotides modulate gene expression during development of adaptive immunity is still not clear.

The immunostimulatory potential of levamisole in fish is of considerable interest in the USA and elsewhere, because it has approval by the U.S. Food and Drug

Administration for treatment of helminthes infections in ruminants. Levamisole a synthetic phenylimidazolthiazole has also been found to be possible modulator of the immune responses of carp and rainbow trout. After treatment with levamisole both fish species showed enhanced nonspecific immune activity and resistance to an experimental challenge with pathogenic bacteria. The increased protection may be correlated with increased phagocytosis, cytokine expression by macrophages, lymphocyte proliferation following exposure to mitogens and antibody response.

The immunostimulants properties of whole microorganisms, chitin particles, lactoferrin, sodium alginates, vitamin C and E, dietary carbohydrates also received considerable attention in fish and shellfish health management strategies. Little information exists regarding the *in vitro* effect of chitin on fish immune system. Some studies do exists on the *in vivo* administration of chitin and it has been found that injection or dietary administration of chitin may enhance the innate immune system of several fish species. Dietary administration of lactoferrin, a glycoprotein enhances the nonspecific immune responses of gilthead seabream Sparus auratus. Sodium alginate extracted from brown algae has been reported to enhance the resistance of common carp Cyprinus carpio against Edwadsiella tarda infection and increase the nonspecific defence system of C. carpio. Alginic acid (Ergosan) and yeast β-glucan (Macrogard) activate innate immune response in sea bass

(Dicentrarchus labrax), particularly under conditions of immunodepression related to environmental stress.

In recent years, whole microorganisms have been tested for their possible immunostimulant properties in fish. The oral administration or injection of yeasts Saccharomyces cerevisiae or Candida utilis and fungus Mucor circinelloides have been shown to increase both humoral and cellular immune responses and to increase or confer resistance against pathogenic bacteria in channel catfish, rainbow trout and gilthead sea bream. Achyaranthes aspera, a herb that stimulates both specific and nonspecific immunity in Indian major carp, Catla catla.

Dietary supplementation of certain vitamins may be effective means of increasing immunocompetency and disease resistance of fish. Elevated doses of vitamin C have been shown to enhance immune responses such as macrophage activities, cell proliferation, natural killer cell activity, complement and lysozyme levels in fish. A positive effect of vitamin C together with vitamin E on the immune response of fish has also been found. Feeding high levels of ascorbic acid has been reported to enhance protection against bacterial infections *viz. Edwardsiella tarda, E. ictaluri, V. anguillarum, A. salmonicida* and against parasitic infection (*Ichthyopthirius multifilis*). High level of dietary vitamin C has been used to counteract immunosuppression caused by aflatoxin B1 contaminated feed in immunocompromised rohu (*Labeo rohita*). Non-gelatinized carbohydrates (46 per cent) along with supplementation with 50 mg kg^{-1} amylase stimulated the immune system in *L. rohita* juveniles. Therefore, immunostimulants have an immense importance in disease management in aquaculture.

Studies on the effect of immunostimulants on shrimp and prawn are still at an infant stage. ß-glucan have received attention in recent years for their ability to increase disease resistance in shrimp and prawn because the limitation of the specific immune response of these animals and the nature of the disease agents make the development of effective vaccine impractical. ß-glucan have been successfully used to increase the resistance of Penaeus japonicus against vibriosis, further studies using Penaeus monodon showed protection against vibriosis, white spot syndrome virus and *Vibrio damsela* and *V. harveyi* and also enhancement of survival and immunity during brood stock rearing. All these effects were caused by direct impact on haemocytes *via* stimulation of phagocytosis, cell adhesion and superoxide anion. Surprisingly the glucan-induced resistance is maternally transmitted. Surprising data has been obtained in the fresh water crayfish *Pacifastatus lenieusculus*. The injection of glucan caused a shortterm severe loss of haemocytes, followed by a rapid recovery due to accelerated release of cells from the haematopoietic organs.

Macrobrachium rosenbergii post larvae showed enhanced growth and resistance to *V. alginolyticus* by dietary administration of ß-glucan. Dietary administration of sodium alginate enhances the immune ability of white shrimp *L. vannamei* and increase its resistance against *V. alginolyticus* infection. Vitamin C has also plays a key role in animal health as an antioxidant by inactivating damaging free radicals producing through normal cellular activity and diverse stressors. Inadequate dietary levels of vitamin C in juvenile shrimp may result in black dead syndrome, reduced growth rates, poor feed conversion ratios, decreased resistance to stress and reduced capability to heal wounds. vitamin C plays a role as immunostimulants, as evidenced by the ability of *P. monodon* postlarvae and juveniles to avoid baculovirus and to resist disease caused by *V. harveyi* and saline shock. However, the mode of action of vitamin C as an immunostimulant is not clear, although its antioxidant role and in consequence cell protection could be a mechanism to preserve haemocytes, improving the general immunological system of shrimp. Therefore results suggest that although the shrimp immune system is nonspecific, it would be possible to enhance disease resistance against pathogens in shrimp by careful and regular use of immunostimulants.

D. Probiotics

The term probiotic means life; it was derived from two Greek words 'pro' and 'bios'. Probiotics are live microbes that can be used to improve the host intestinal microbial balance and growth performance. Development of probiotics in aquaculture management will reduce the use of antimicrobial drugs which were prophylactive alone and whose over dependence in recent times poses potential hazards to man who consume them.

Selection of Probiotics

The initial major purpose of using probiotics is to maintain or reestablish a favorable relationship between friendly and pathogenic microorganism that constitute the flora of intestinal or skin mucus of fish. A successful probiotic is expected to have a few specific properties in order to certify, a beneficial effect.

A healthy source of microorganisms from a digestive tract of healthy aquatic animals must be selected. The microorganisms with which the work is to be carried out are isolated and identified by means of selective culture. A new culture with only the colonies of interest for conducting *in vitro* evaluations such as inhibition of pathogens; pathogenicity to target species; resistance conditions of host; among others are performed. In case of the absence of restrictions on the use of the target species, experiments with *in vivo* supplementation, and small and large scale, are carried out to check if there are real benefits to the host.

Finally, the probiotic that presented significantly satisfactory result can be produced commercially and utilized.

Selection of Candidate Species

Generally at the end of the previous phase, one ends up with a pool of isolates that must be screened and preselected to obtain a restricted number of isolates for further examination. Several approaches followed in literature are discussed below. The application of *in vitro* tests to screen the acquired bacterial strains presupposes a well-known mode of action to select the appropriate *in vitro* test.

Since the evidence about the possible modes of action of probiotics is still equivocal, preference should be given to *in vivo* tests in the search for probiotics. The use of the target organism in the screening procedure provides a stronger basis.

In vitro antagonism testes a common way to screen the candidate probiotics. To perform *in vitro* antagonism test pathogens are exposed to the candidate probiotics or their extracellular products in liquid (Gildberg *et al.,* 1995) or solid (Dopazo *et al.,* 1988) medium. Depending on the exact arrangement of the tests, candidate probiotics can be selected based on the production of inhibitory compounds or siderophores, or on the competition for nutrients.

Colonization and adhesion: It is stated above that a candidate probiotic should either be supplied on a regular basis or be able to colonize and persist in the host or in its ambient environment. The ability of a strain to colonize the gut or an external surface of the host and adhere to the mucus layer may be a good criterion for preselection among the putative probiotics. This involves the viability of the potential probiotics within the host and/or within its culture environment, adherence to host surfaces, and the ability to prevent the establishment of potentially pathogenic bacteria. Examination of adhesion properties using intestinal cells has become a standard procedure for selecting new probiotic strains for human application (Salminen *et al.,* 1996), but it is less common in aquaculture (Jöborn *et al.,* 1997).

Evaluation of Pathogenicity of Selected Strains

Before a culture can be used as a probiotic, it is necessary to confirm that no pathogenic effects can occur in the host. Therefore, the target species should be challenged with the candidate probiotic, under normal or stress conditions. This can be done by injection challenges, by bathing the host in a suspension of the candidate probiotic, or by adding the probiotic to the culture. Garriques *et al.* (1995) examined the pathogenicity of three bacterial isolates toward *Litopenaeus vannamei* by adding these to the nauplii cultures. Mortality was recorded for up to 4 days, and the animals were monitored for phototactic response.

In vivo Evaluation of Potential Probiotic Effects on the Host

The effect of candidate probiotics should be tested *in vivo* as well. When the probiotic effect is supposed to be nutritional, the candidate probiotics could be added to the culture of the aquatic species and their effect on growth and/or survival parameters could be assessed. However, when biological control of the microbiota is desired, representative *in vivo* challenge tests seem to be the appropriate tool to evaluate the potential effect of the candidate probiotic on the host.

Mode of Application of the Putative Probiotic

The putative probiotic can be added to the host or its ambient environment in several ways: (i) addition to the artificial diet, (ii) addition to the culture water, (iii) bathing and (iv) addition *via* live food.

Experimental Infections–*In vivo* Antagonism Tests

The next important step in the *in vivo* challenge test involves experimental infection with a representative pathogen. Pathogens or opportunistic pathogens can be administered *via* the diet (live food or artificial), through immersion, or *via* the culture water, similarly to the probiotics. In experiments performed by Gildberg *et al.*, Atlantic salmon fry was challenged with cohabitant fishes which were previously infected with *Aeromonas salmonicida* through intraperitoneal injection. The mortality of the challenged fishes due to acute furunculosis was monitored over time. *In vivo* challenge tests to detect probiotic effects should be performed with great caution. The long-term effects should be studied to determine how the pathogen is evolving. It is important to observe whether the growth or the activity of the pathogen in the location where the antagonism of the probiotics is expected is really suppressed or if it is only delayed, may be simply due to some competition for nutrients.

Mode of Action

Those modes are as follows: production of inhibitory compounds, competition for chemicals or available energy, competition for adhesion sites, enhancement of the immune response, improvement of water quality, interaction with phytoplankton, source of macro- and micronutrients and enzymatic contribution to digestion. The possible modes of action are discussed in more detail in the following sections.

a. Competitive Exclusion

i. Competition for Nutrients

Competition for nutrients can theoretically play an important role in the composition of the microbiota of the intestinal tract or ambient environment of cultured aquatic species. Hence, successful application of the principles of competition to natural situations is not easy and remains a major task for microbial ecologists. The microbial ecosystem in aquaculture environments is generally dominated by heterotrophs competing for organic substrates as both carbon and energy sources. Specific knowledge of the factors governing the composition of the microbiota in aquaculture systems is required to manipulate it. Nevertheless, Rico- Mora *et al.* (1998) selected a bacterial strain for its active growth in organic-poor substrates and inoculated it into a diatom culture, where it prevented the establishment of an introduced *V. alginolyticus* strain. Since the inoculated strain had no *in vitro* inhibitory effect on *V. alginolyticus,* it was suggested that the strain was able to outcompete *V. alginolyticus* due to its ability to utilize the exudates of the diatom. Verschuere *et al.,* selected several strains with a positive effect on the survival and growth of Artemia juveniles.

ii. Competition for Adhesion Site

One possible mechanism for preventing colonization by pathogens is competition for adhesion sites on gut or other tissue surfaces. It is known that the ability to adhere to enteric mucus and wall surfaces is necessary for bacteria to become established in fish intestines. Since bacterial adhesion to tissue surface is important during the initial stages of pathogenic infection, competition for adhesion receptors with pathogens might be the first probiotic effect. Adhesion capacity and growth on or in intestinal or external mucus has been demonstrated *in vitro* for fish pathogens like *V. anguillarum* and *A. hydrophila* and for candidate probiotics such as *Carnobacterium* strain K1 and several isolates inhibitory to *V. anguillarum.* The intestinal isolates generally adhered much better to a film of turbot intestinal mucus, skin mucus, and bovine serum albumin than did *V. anguillarum,* indicating that they could compete effectively with the pathogen for adhesion sites on the mucosal intestinal surface.

iii. Antibiotics Production

Colonization of the fish digestive tract by bacteria capable of producing lactic acid fermentation may inhibit the proliferation of putrefactive microbes in that site, thus protecting the host from diseases caused by toxins generated by proteolytic bacteria. Bacteriocins are bactericidal or bacteriostatic peptides that are mostly active against bacteria closely related to the producer. Nisin, produced by some *L. lactis* strains, is by far the best known and most studied.

The carnocin from *C. piscicola* was also effective against *A. hydrophila,* though to a lesser extent than against *L. monocytogenes.* The inhibitory effect of lactic acid bacteria against fish pathogen is not limited to strains isolated in fish. Many lactic acid bacteria produce inhibitory compounds against *A. hydrophila* (Lewus *et al.,* 1991; Santos *et al.,* 1996; Gopalakannan *et al.,* 2003). The fact that fish have a gut flora with inhibitory effects against pathogens may have relevance to fish health, and further studies are therefore needed.

iv. Siderophores Production

Virtually all microorganisms require iron for growth. Siderophores are low-molecular-weight (1,500), ferric ion-specific chelating agents which can dissolve precipitated iron and make it available for microbial growth. The ecological significance of siderophores resides in their capacity to scavenge an essential nutrient from the environment and deprive competitors of it. Successful bacterial pathogens are able to compete successfully for iron in the highly iron-stressed environment of the tissues and body fluids of the host. The importance and the biological role of ferric iron in the establishment of the microbiota associated with cultured aquatic species. Harmless bacteria which can produce siderophores could be used as probiotics to compete with pathogens whose pathogenicity is known to be due to siderophore production and competition for iron or to outcompete all kind of organisms requiring ferric iron from solution.

Characteristics of Good Probiotics

The following are the features of good probiotic bacteria:

1. It should be a strain, which is capable of exerting a beneficial effect on the host animal *e.g.* increased growth or resistance to disease.
2. It should be non-pathogenic and non-toxic.
3. It should be present as viable cells preferable in large numbers.
4. It should be capable of surviving and metabolizing in the gut environment *e.g.* resistance to low pH and organic acid.
5. It should be stable and capable of remaining viable for periods under storage and field conditions.

A probiotic agent with all these features has considerable advantage over antibacterial supplements such as antibiotics currently in use. They do not induce resistance to antibiotics which will compromise therapy. They are not toxic and therefore will not produce undesirable side effect when being fed and in the case of food animal will not produce toxic residues in the carcass. They may stimulate immunity whereas the immune status remains unaffected by antibiotics.

Constraints to Probiotics in Aquaculture

1. Inability of strains to be produced in commercial quantities and consequent demonstration on a large scale.
2. Difficulty in proving performance at the farm level.
3. Inability of companies to conduct extensive research on how to make product specifically for aquaculture purposes.

Probiotics Significance in Aquaculture

There are some possible benefits linked to the administering of probiotics which have already been suggested as:

Improvement in Water Qualities

Nitrogenous compounds contamination such as ammonia, nitrite and nitrate in fish culture systems/ponds has been a serious concern. The susceptibility of cultured aquatic species to high concentration of these compounds is generally species-specific, but in high concentrations, these compounds may be extremely harmful and cause mass mortality in all cases. The ability of *Lactobacillus* spp. JK-8 and JK-11 simultaneously removes nitrogen and pathogens from contaminated shrimp farms. In several other studies, water quality has been improved by the addition of probiotics especially *Bacillus* spp. The reason is that gram – positive *Bacillus* spp. generally more efficient in converting organic matter back to CO_2 than gram – negative bacteria, which would convert a greater percentage of organic carbon to bacterial biomass or slime.

As Growth Promoters

It has been demonstrated experimentally that probiotics indeed may enhance the growth of fish. The ability of organisms to out-grow the pathogens in favour of host or to improve the growth of the host and yet no side effect on the host made it a probiotic bacteria. probiotic bacteria as growth promoter on tilapia (*Oreochromis niloticus*) identified that the highest growth performance was recorded with *Micrococcus luteus* a probiotic and the best feed conversion ratio was observed with the same organism. So *M. luteus* may be considered as a growth promoters in fish aquaculture. Lactic acid bacteria also had an effect as growth promoters on the growth rate in juvenile carp though not in Sea bass.

For Disease Prevention

Probiotics or their products for health benefits to the host have been found useful in aquaculture, terrestrial animals and in human disease control. These include microbial adjunct that prevent pathogens from proliferating in the intestinal tract, on the superficial surfaces and in culture environment of the culture species. The effect of these beneficial organisms is achieved through optimizing the immune system of culture organism, increasing their resistance to disease, or producing inhibitory-substance that prevent the pathogenic organisms from establishing disease in the host.

Source of Nutrients and Enzymatic Contribution to Digestion

Some researchers have suggested that microorganisms have a beneficial effect in the digestive processes of aquatic animals. In fish, it has been reported that Bacteroides and *Clostridium* sp. have contributed to the host's nutrition, especially by supplying fatty acids and vitamins. Some microorganisms such as *Agrobacterium* sp., *Pseudomonas* sp., *Brevibacterium* sp., *Microbacterium* sp., and *Staphylococcus* sp. may contribute to nutritional processes in *Salvelinus alpinus*. Microbiota may serve as a supplementary source of food and microbial activity in the digestive tract may be a source of vitamins, or essential amino acids

Stress Tolerance

Intensive production of aquaculture results in stress in fishes. Zebra fish and European seabass stress were reduced by food supplemented with probiotics bacteria *Lactobacillus delbruecki* sp. Another way to assess stress in fish involves subjecting them to heat shock. In this method of stress reduced in Japanese flounder while fed with probiotics incorporated feed. The gill head seabream (*Sparus auratus*) fed with probiotics *Alteromonas* sp. showed decreased stress indicator levels of lactate and plasma glucose level.

Enhancement of the Immune Response

Among the numerous beneficial effects of probiotics, modulation of immune system is one of the most commonly purported benefits of probiotics. Fish larvae, shrimps and other invertebrates have immune systems that are less well developed than adult stage and are dependent primarily on non-specific immune responses for their resistance to infection evaluated the ability of *Lactobacillus fermentum* LbFF4 isolated from Nigerian fermented food ('fufu') and *L. plantarum* LbOGI from a beverage 'Ogi' to induce immunity in *Clarias gariepinus* (Burchell) against some selected fish bacterial pathogens.

Antiviral Effect

Some probiotics bacteria have the properties of an antiviral effect. Strains of *Pseudomonas* sp., *Vibriosis* sp., *Aeromonas* sp and groups of Cornyforms isolated from Salmonids hatcheries showed the aniviral activity against infectious hematopoitic necrosis virus with more than 50 per cent plaque reduction. Like vise *Vibrio* sp. NICA1030 and NICA1031 isolated from black tiger shrimp sowed antiviral activity against IHNV and onchorynchus masou virus with percentage of plaque reduction between 62 and 99 per cent respectively. The isolation of a strain of pseudo alteromonas undine which exerted antiviral effects by increased survival in shrimp which was experimentally infected with simaaji necrosis virus, baculovirus and irido virus.

Effect on Reproduction of Aquatic Species

The effect of probiotics supplementation on reproductive performance of fish was studied using *B. subtilis* isolated from intestine of *Cirrhinus mirigala*. Probiotic incorporated at different concentration to fed species of ornamental fishes *Poeciliareticulate, P.sphenops,Xiphophorus heleri* and *X. maculates* in a one year experiment. In that, *B. subtilis* concentration of 10^6-10^8 cells per g of feed produced increase the gonodosomatic index, fecundity viability and production of fry from the females of all four species and also those probiotics bacteria were synthesized the complex B vitamins such as thiamine (vit B1) and Vit B2 which was helpful to reduce the number of dead or deformed alevins.

Probiotics in Aquaculture Management

These organisms can be administered to the aquaculture management through feeding, injection or immersion of the probiotic bacteria.

Application in Feed

Probiotics are applied with the feed and a binder (egg or cod liver oil) and most commercial preparation contain either *Lactobacillus* sp. or *Saccharomyces cerevisiae*. According to FAO and WHO guidelines, probiotic organisms used in food must be capable of surviving passages through the gut *i.e.* they must have the ability to resist gastric juices and exposure to bile. Furthermore they must be able to proliferate and colonize the digestive tract and they must be safe, effective and maintain their effectiveness and potency for the duration of the shelf life of the product.

Direct Application to Pond Water

The water probioics contain multiple strains of bacteria like *Bacillus acidophilus, B. subtilis, B. lecheniformis, Nitrobacter* sp. *Aerobacter* and *Sacharomyces cerevisiae*.

Application of probiotic through water of tanks and ponds may also have an effect on fish health by improving several qualities of water, since they modify the bacteria composition of the water and sediments.

Application through Injection

Application of probiotics by injection is a possibility. The possibility of freeze-drying the probiont like vaccine and applied either through bathing, or injection. It has demonstrated the experimental administration of probiotic *Micrococcus luteus* to *Oreochromis niloticus* by injection through intra peritoneal route which had only 25 per cent mortality as against 90 per cent with *Pseudomonas* using the same route. The use of probiotics stimulate Rainbow trout immunity by stimulating phagocytes activity, complement mediated bacterial killing and immunoglobulin production.

Bioremediators and Other Prophylactic Measures Bioremediators

Environmental probionts or bioremediators improve water quality by nitrification, removal of toxic substances such as ammonia, hydrogen sulphide and nitrite from the culture systems. They would efficiently recycle the waste material by decomposing the complex organic substances and generating nutrients in the form of simple inorganic compounds. Of the several strains of bacteria, gram-positive *Bacillus* spp. are generally more efficient in converting organic matter back to CO_2 than are gram-negative bacteria. Other bacterial species of importance towards bioremediation are: *Nitrobacter, Pseudomonas, Enterobacter, Cellulomonas,* and *Rhodopseudomonas* spp. Bioremediators (biocontrol agents) apart from improving water quality parameters, modify or manipulate the inherent microbial communities and reduce the pathogenic microbes. The cell densities of these organisms are greatly influenced by the carbon-nitrogen ratio of the sediment and that addition of carbon source could increase the proliferation of heterotrophic bacteria compared to the pathogenic bacteria. Build up of ammonia in the culture systems are removed by nitrifiers by oxidation of ammonia to nitrite and subsequently to nitrate.

Other Prophylactic Measures

Other prophylactic measures include management aspects aimed at controlling disease incidences. Selection of healthy juveniles and avoiding contamination during culture is an effective step for disease control. Specifically, quarantining the stocks that are imported or transferred over a wide geographic area has to be practised. Disease avoidance by using specific pathogen free (SPF) stock is a very useful method for disease prevention. Development and raising of specific pathogen resistant (SPR) strains through selective breeding of animals is one of the potential approaches, which has immense use in the broodstock management programmes.

Use of biosecure (preventing entry of infected host and carriers), closed or semi - closed recycle system with reduced water exchange and increased water-reuse culture systems, disinfection of contaminated water before discharge in case of an outbreak, maintaining optimum stocking density, use of virus-free broodstock to prevent vertical transmission of the viral diseases, avoiding too much organic loading and regular monitoring the health status of the cultured animals by suitable checking methods are the recommended measures for a disease free aquaculture system.

Despite the best biological and management measures undertaken for controlling the diseases, prevention of disease outbreaks sometimes become impossible in an aquaculture farm or hatcheries. In such conditions, chemotherapy would become essential with chemotherapeutants that are eco-friendly and less toxic. Antibiotic application has be done with caution in controlled condition after identifying the correct antibiotic through antibiogram assay. Other chemical disinfectants like formalin, chelated copper compounds, quaternary ammonium compounds (benzalkonium chloride – BKC), bleaching powder and potassium permanganate are used at appropriate concentrations and in a rational manner for treating specific cases of infection.

Glossary of Fish Health Terms

Abrasion: a localized area denuded ill skin, mucous membranes, or superficial epithelium caused by- rubbing or scraping.

Abscess: a localized inflammation and swelling, frequently filled with necrotic debris and white blood cells.

Acclimation: the process through which fish become fully adapted: to new environmental circumstances; such as being placed into water of different quality, temperature, or different holding situations.

Acid fast: bacteria that retain red phenolic fuchsin stain after being treated with add alcohol solution.

Acute: severe or crudal, often progressing rapidly; *i.e.*, acute inflammation.

Adhesion: the abnormal fibrous union of an organ or part to another.

Adjuvant: a material administered with and enhancing the action of a drug or antigenic substance.

Adipose (tissue): fatty animal tissue.

Aerobic: said of an organism or life process that utilizes or can only ©cist in the presence of oxygen.

Anadromous: fish that leave the sea and migrate to fresh water to spawn.

Anaerobic: said of an organism or life process that flourishes in the absence of oxygen.

Anemia: a condition characterized by a deficiency of hemoglobin or red blood cells (erythrocytes).

Aneurysm: a sac formed by the dilation of the walls of an artery or a vein and filled with blood.

Anthelminthic: an agent that destroys: or expels parasitic worms in the gut.

Antibiotic: a chemical substance produced by living organisms, usually mold or bacteria, that is capable of inhibiting other organisms.

Antibody: a specific immunoglobulin molecule produced by an organism in response to an antigen.

Antigen: a high molecular weight protein or polysaccharide which stimulates the formation of specific antibody with which it will react. Examples include killed bacterial cells or flagella.

Bacteremia: the presence of living bacteria in the blood, with or without significant response on the part of the host: usually refers to a generalized bacterial infection in the blood.

Bacterin: a vaccine prepared from bacteria that have been inactivated by heat or chemicals without altering the cell antigens.

Bacteriocidal: having the ability to kill bacteria.

Bacteriostatic: having the abiity to inhibit or retard the growth or reproduction of bacteria.

Benign: not endangering life or health.

Boil: a furuncle; a localized infection or abscess within subcutaneous tissue that drains externally.

Carcinogen: any agent or substance which produces cancer or accelerates the development of cancer.

Carrier: an individual harboring the specific organism(s) which can cause a disease, without indication of signs of the disease.

Catadromous: fish that leave fresh water and migrate to the sea to spawn.

Cataract: partial or complete opacity of the crystalline lens of the eye or its capsule.

Chemotherapeutic: a chemical agent used for the prevention or treatment of disease.

Cilia: short hair-like processes on protozoans by which they move or produce currents.

Clinical: when applied to a disease or signs of disease, a term that indicates a condition is readily apparent, overt, or obvious by gross inspection.

Coagulation: the process of clotting.

Communicable disease: a disease that is naturally transmitted directly or indirectly from one individual to another.

Complement: factors present in the serum of normal animals which enter into various immunologic reactions.

Culture: population of bacteria grown on artificial medium.

Culture media: material (solid or liquid) on which bacteria are grown.

Disease: a pathological condition of the body that presents a group of signs indicating the existence of an abnormal histological or physiological entity,

Disinfectant: an agent which will destroy infective agents.

Ectoparasite: a parasite that lives on the external surface of the host.

Edema: excessive accumulation of fluid in the tissue space or body cavities.

Embolus: undissolved material carried in the bloodstream, such as a blood clot, air bubbles, cancerous or other tissue cells, fat, clumps of bacteria, or a foreign body.

Endogenous: originating in the cells or tissues of the body.

Endoparasite: a parasite that lives within the host.

Enteritis: any inflammation of the intestinal tract.

Enzootic: a disease which is present in an animal population at all times.

Epizootic: outbreak of disease attacking many animals in a population at the same time and rapidly spreading.

Etiology: the study of the causes of a disease.

Exophthalmos: abnormal protrusion of the eyeball from the socket.

Facultative fish pathogens: occurring naturally as non-pathogens in the environment but capable of causing disease outbreaks under conditions of stress.

Flagella: whip-lite organelles of locomotion on protozoans.

Free-Living: not requiring a host to survive.

Furuncle: a localized infection of skin or subcutaneous tissue which develops a solitary abscess that may or may not drain externally.

Gram-negative: bacteria which lose the purple crystal violet stain when treated with alcohol solution in the Gram- staining process.

Gram-positive: bacteria which retain the purple crystal violet stain when treated with alcohol solution in the Gram-staining process.

Gross pathology: pathology that deals with the superficial or overt appearance of organs and tissues.

Hematocrit: volumetric relationship of the cellular elements of blood to the total blood volume; sometimes referred to as the packed cell volume.

Hemoglobin: the respiratory pigment of erythrocytes capable of taking up and giving off oxygen.

Hemolysis: detraction of erythrocytes.

Hemorrhage: an escape of blood from the vessels, either through intact blood vessel walls or through ruptured vessels.

Histopathology: the study of microscopic changes in diseased tissue

Host: an animal or plant which harbors or nourishes another organism.

Hyper: a prefix denoting excessive, above normal or situated above.

Hyperplasia: abnormal increase in the number of cells in a tissue or organ accompanied by enlargement or an increase in the size of the tissue or organ.

Hypertrophy: enlargement of an organ due to an increase in the size of cells rather than in the number of cells.

Hypo: a prefix denoting a deficiency, less than normal, below or beneath.

Immunity: resistance to disease; lack of susceptibility.

Immunization: the act or process of rendering immune by the introduction or administration of an antigen.

Incubation: period of time between exposure or introduction of pathogens into the host and development of typical signs of disease.

Inflammation: the reaction of the tissues to infection or injury characterized clinically by swelling and redness.

Inoculation: the introduction of a pathogenic organism into the tissues of a living organism or culture medium.

Intra: within or between layers of same tissue.

In vitro: used in reference to tests or experiments conducted in vessels or in an artificial environment.

In vivo: used in reference to tests or experiments conducted in or on living organisms.

-itis: a suffix indicating inflammation.

Lesion: any visible alteration in the normal structure of organs, tissues, or cells.

Lordosis: dorso-ventral curvature of the spine.

Lysozyme: an enzyme which is capable of destroying certain bacterial cell walls.

Melanin: a dark pigment responsible for the yellow to black coloration of fishes.

Moribund: obviously progressing towards death, nearly dead.

Morphology: the study of the form and structure of animals and plants.

Mortality: the death rate, also the ratio of dead to living individuals in a population.

Mucus: the slime produced by mucous membranes or by special cells in fish skin.

Necropsy: a medical examination of animals to ascertain the cause of death.

Necrosis: the process of death of cells or tissues within the living body.

Non-pathogenic: refers to an organism which does not cause disease.

Obligate fish pathogens: disease-causing organisms that cannot survive in nature unless susceptible or carrier fish are present.

-oma: a suffix used to denote tumours, *i.e.* fibroma.

Opportunistic pathogen: an organism capable of causing disease only when the host's resistance is lowered or when unusual circumstances favor its growth and development.

Overt disease: a disease, not necessarily infectious, that is apparent or obvious by gross inspection; a disease exhibiting obvious clinical signs.

Parasite: an organism that lives in or on another organism (the host), that depends on the host for its food, and that is suspected of harming the host when present in large numbers.

Pathogenesis: the origin and process of development of any disease or morbid process.

Pathogenic: causing disease.

Petechia: a minute hemorrhage on a surface.

Predisposing factors: physical, chemical or biological factors which increase the susceptibility of an organism (host) to disease; sometimes called stressors.

Prophylaxis: actions taken to prevent disease or measures taken to prevent the development or spread of disease.

Putrefaction: the enzymatic decomposition of organic matter, especially proteins, by anaerobic micro- organisms.

Resistance: a natural ability of an organism to withstand the effects of various physical, chemical, and biological agents which might otherwise cause disease in the organism.

Sensitive, drug: susceptibility of a micro-organism, usually a bacterium, to be controlled (inhibited or killed by use of a drug).

Septicemia: generally involving the significant invasion of the bloodstream by micro-organisms; a severe bacterial infection in the blood.

Sign: any manifestation of disease, such as an aberration in structure, physiology, appearance or behaviour, as interpreted by an observer.

Specificity, host: extreme host requirements that limit a parasite to one host species only. Loose host specificity indicates a parasite can infect many hosts.

Subcutaneous: beneath the skin.

Synergism: refers to an interaction wherein two agents produce a greater effect than would be predicted from the sum of their individual effects.

Scoliosis: lateral curvature of the spine.

Therapeutic: serving to heal or cure.

Toxicity: ability of a substance to kill or cause an adverse effect.

Ubiquitous: universally or widely distributed.

Ulcer: an open sore or a break in the skin or a mucous membrane with loss of surface tissue; localized disintegration and necrosis of epithelial tissue.

Vaccine: a preparation of nonvirulent or killed disease organisms administered into the body to stimuate the production of antibodies against them.

Vector: a living organism which carries an infectious agent from one infected individual to another, directly or indirectly.

Virulence: the relative capability of a pathogen to produce disease.

References

Alan C. 2005. Molecular Virology. Academic Press.

Andrews C, Excell A and Carrington N. 1988. The Manual of Fish Health. Salamander Books Ltd.

Austin B and Austin DA. 1993. Bacterial Fish Pathogens. Disease in Farmed and Wild Fish. 2nd Ed. Ellis Horwood.

David SA, Lee CS and O'Bryen PJ. 2006. Aquaculture Biosecurity- Prevention, Control and Eradication of Aquatic Animal Diseases. World Aquaculture Society. Blackwell.

David MK, Peter MH, Diane EG, Robert AL, Malcolm AM, Bernard R and Stephen ES. 2007. Fields Virology. 5th Ed.Lippincott Williams and Wilkins.

Freshney IR. 2005. Culture of Animal Cells: A Manual of Basic Technique. 3rd Ed. John Wiley and Sons.

Eiras J, Segner H, Wahli T and Kapoor BG. 2008. Fish Diseases. Science Publ.

Ferguson HW. (Ed). 2006. Systemic Pathology of Fish: A Text and Atlas of Normal Tissues in Teleosts and their Responses in Disease. 2nd Ed. Scotian Press.

Felix S, Riji John K, Prince Jeyaseelan MJ and Sundararaj V. 2001. Fish Disease Diagnosis and Health Management. Fisheries College and Research Institute, T.N. Veterinary and Animal Sciences University. Thoothukkudi.

Ferguson HW. (Ed). 2006. Systemic Pathology of Fish: A Text and Atlas of Normal Tissues in Teleosts and their Responses in Disease. 2nd Ed. Scotian Press.

Humphrey J, Arthur JR, Subasinghe RP and Phillips MJ. 2005. Aquatic Animal Quarantine and Health Certification in Asia. FAO.

Inglis V, Roberts RJ and Bromage NR. 1993. Bacterial Diseases of Fish. Blackwell.

John P. 1999. Health Maintenance and Principal Microbial Diseases of Cultured Fishes. 2nd Ed. Blackwell.

Leatherland JF and Woo PTK. 1998. Fish Diseases and Disorders. Vol. II. Non-Infectious Diseases. CABI.

Lom J and Dykova I. 1992. Protozoan Parasites of Fishes. Elsevier.

Noga EJ. 1996. Fish Disease Diagnosis and Treatment. Mosby-Year Book.

Mothersill C and Austin B. 2000. Aquatic Invertebrate Cell Culture. Springer Praxis.

Poulin R and Grimes LR. 2007. Evolutionary Ecology of Parasites. Princeton University Press.

Rhode K. 2005. Marine Parasitology. Steven Simpson Books.

Roberts RJ. 2001. Fish Pathology. 3nd Ed. W.B. Saunders.

Shankar KM and Mohan CV. 2002. Fish and Shellfish Health Management. UNESCO.

Sindermann CJ. 1990. Principal Diseases of Marine Fish and Shellfish. Vols. I, II. 2nd Ed. Academic Press.

Stoskopf MK. 1993. Fish Medicine. WB Saunders.

Wolf K. 1988. Fish Viruses and Viral Diseases. Cornell University Press.

Woo PTK and Bruno DW. (Eds.). 1999. Fish Diseases and Disorders. CABI.

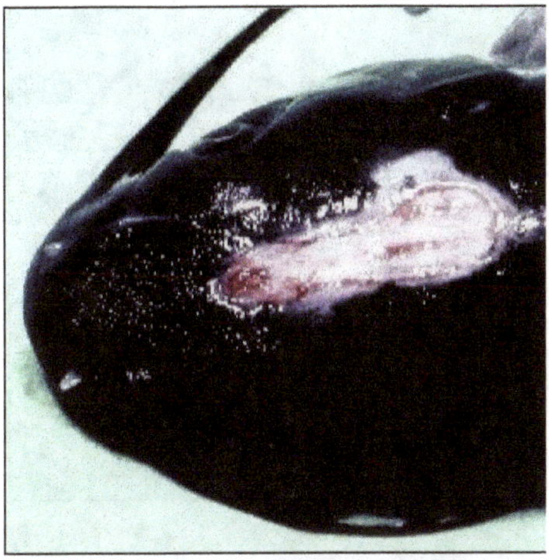

p. 16

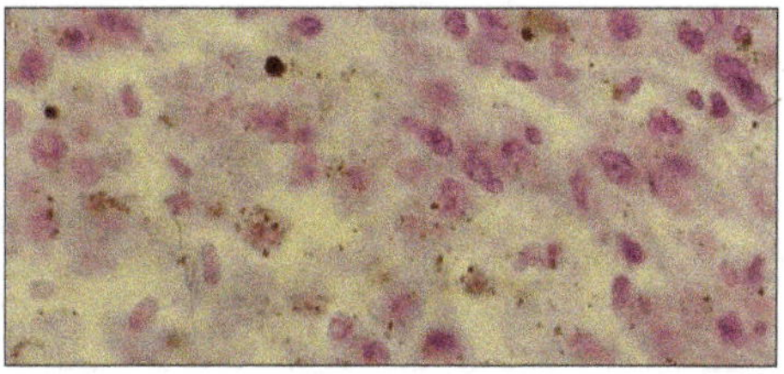

p. 12

p. 17

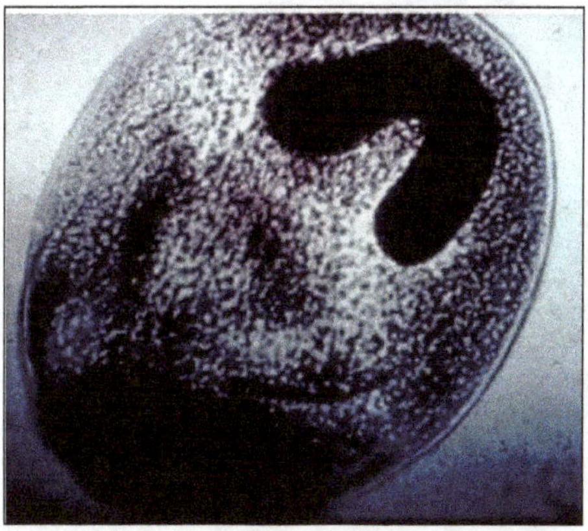

p. 39

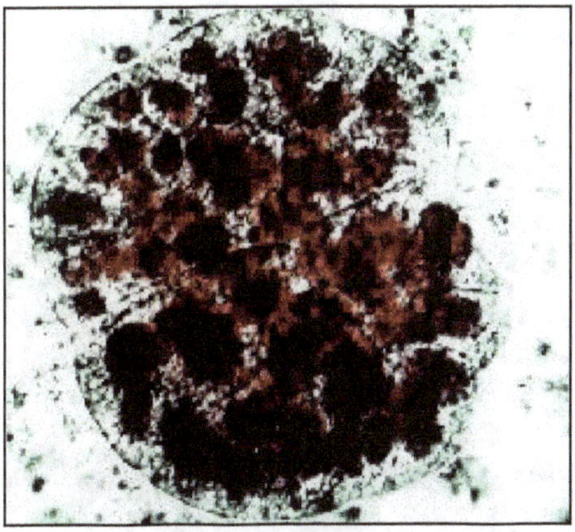

p. 40

p. 40

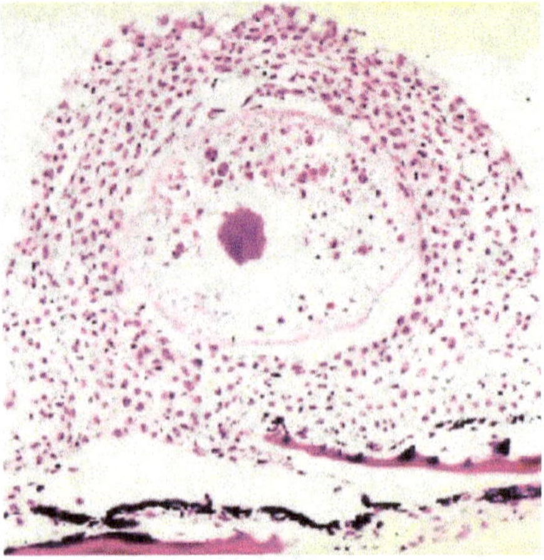

p. 40

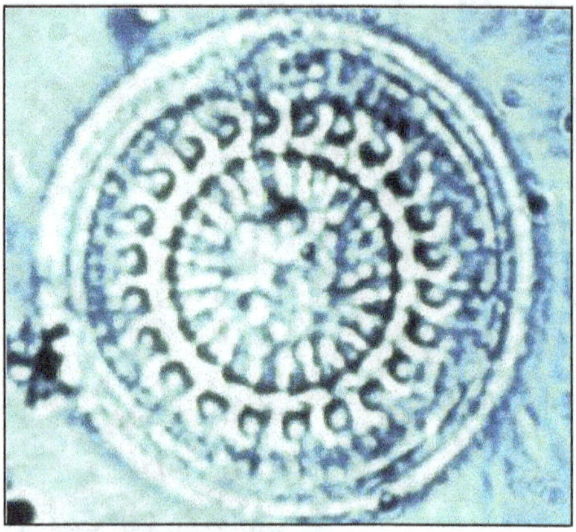

p. 40

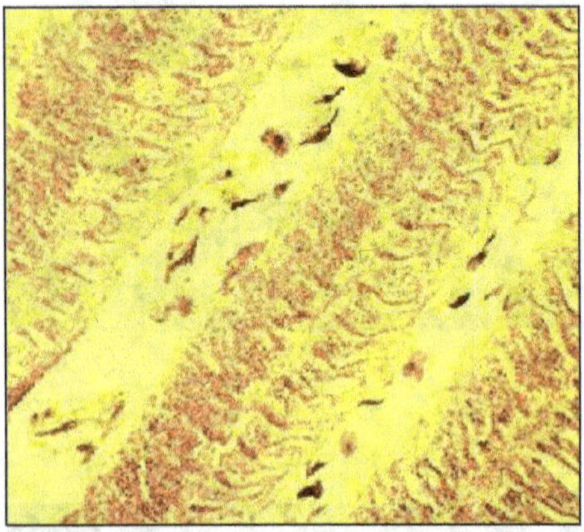

p. 42

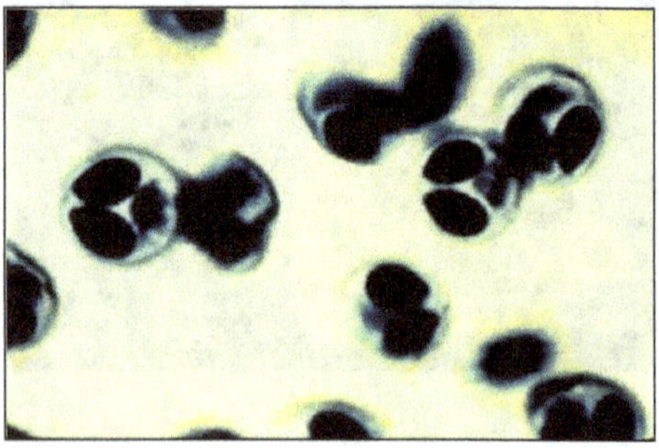

p. 42

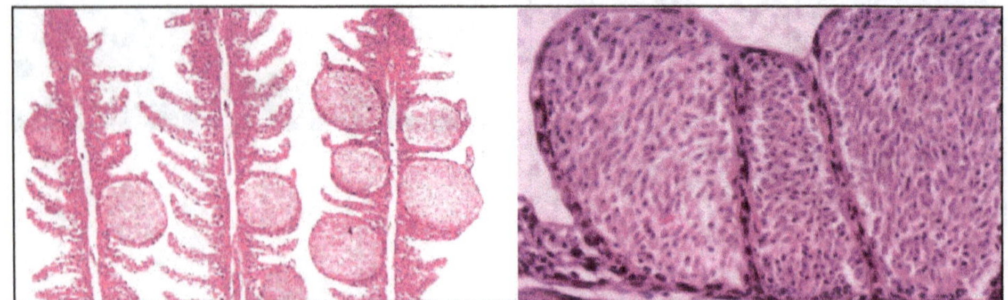

p. 43

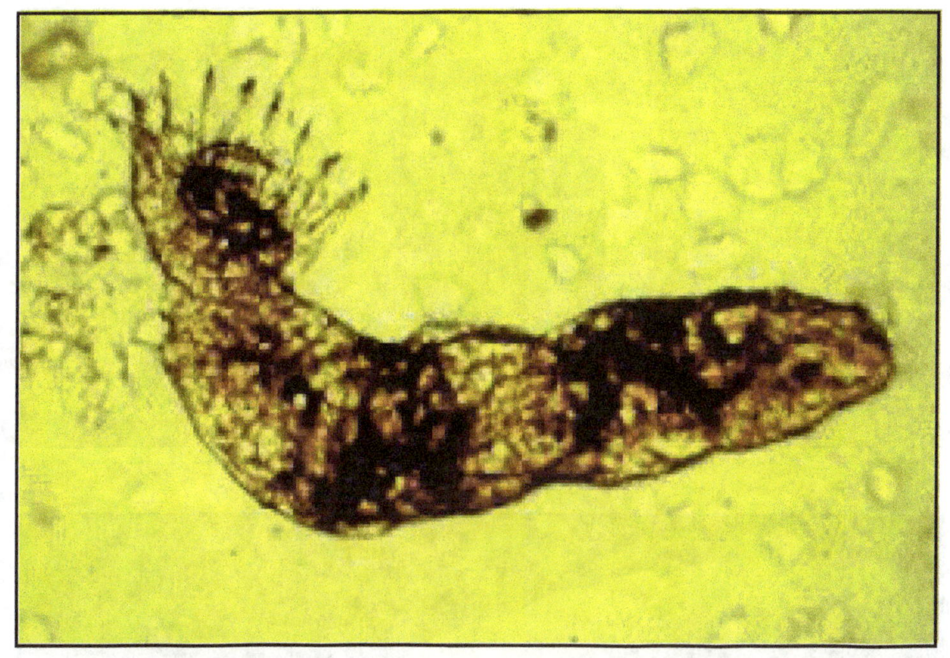

p. 44

p. 45

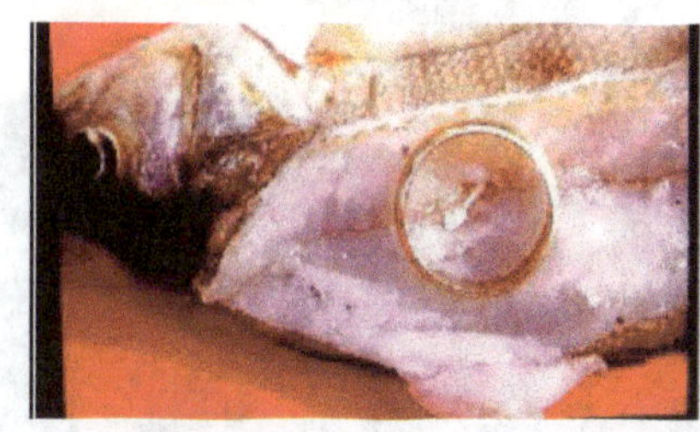

p. 45

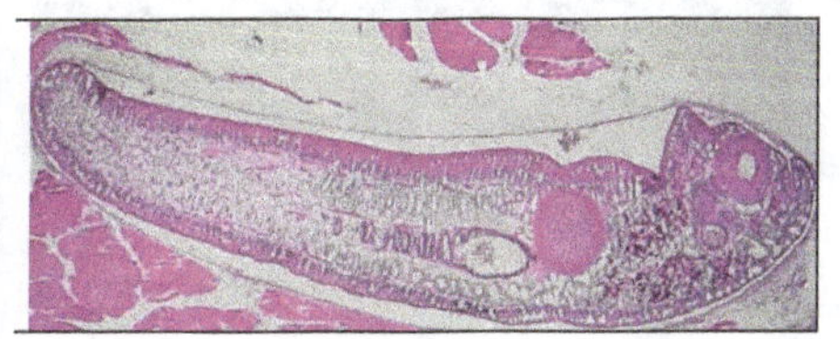

p. 45

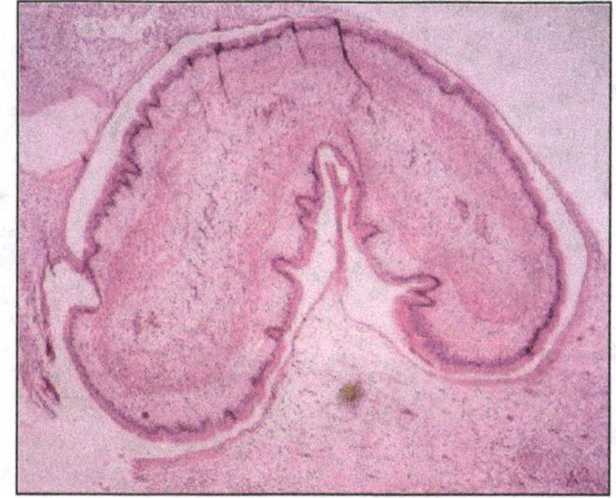

p. 46

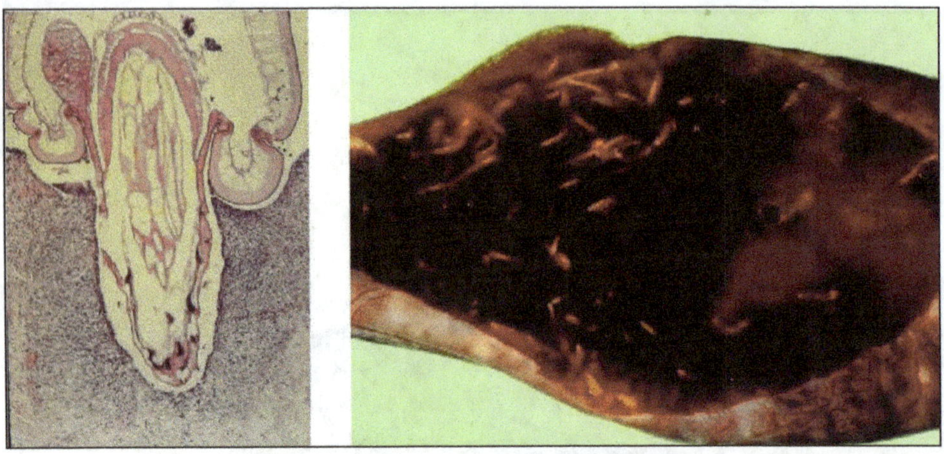

p. 47

p. 47

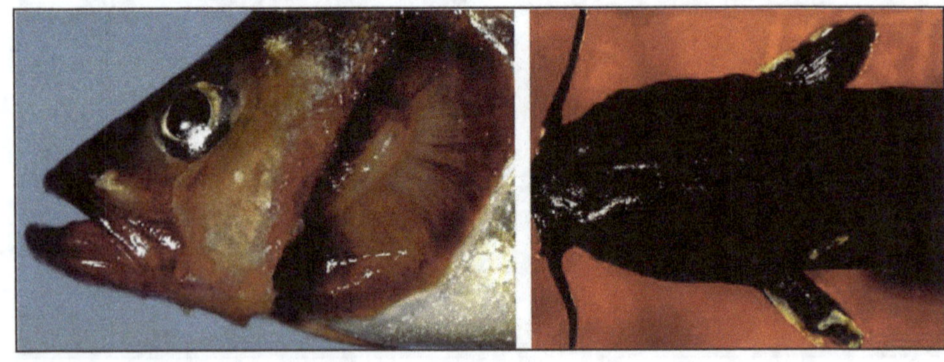

p. 52

www.ingramcontent.com/pod-product-compliance
Lightning Source LLC
LaVergne TN
LVHW012124060726
842759LV00003B/25